COCONUT 24/7

COCONUT 24/7

Easy Ways to Look and Feel Better

Pat Crocker

Collins

Published by HarperCollins Publishers Ltd

First edition

Food styling by Dennis Wood. Prop styling by Susan Rogers (www.rogersdesign.ca).

HarperCollins Publishers Ltd
2 Bloor Street East, 20th Floor
Toronto, Ontario, Canada, M4W 1A8

www.harpercollins.ca

Library and Archives Canada Cataloguing in Publication
information is available upon request

ISBN 978-1-44343-054-8

Printed and bound in Canada
TC 9 8 7 6 5 4 3 2 1

To the coconut, that majestic and health-giving tropical plant, the tower of life itself, and to the 12,000 or so islands—the "coconut isles"—lands of abundance nestled in the vast basin of the great Pacific Ocean between the Tropic of Cancer and the Tropic of Capricorn.

CONTENTS

PREFACE

The coconut has been sharing its gifts of fibre, wood, meat and life-giving water from a time as old as the islands themselves. Antonio Pigafetta, the Italian nobleman who sailed with Magellan, wrote in his journal, "With two of these palm trees, a whole family of ten can sustain itself." Indeed, even the 4,000-year-old Sanskrit words for coconut, *Kalpa Vriksha*, are thought to mean "the tree that gives all that is necessary for living."

Now, thanks to research by nutritionist Mary Enig, Ph.D., Bruce Fife, C.N., N.D., and others, as well as new extraction techniques that offer an array of coconut food products to cooks the world over, we can all enjoy the miracle of coconut.

This book begins with an overview of the incredible healing gifts of coconut. It then explores both familiar and innovative coconut ingredients. It provides reliable, science-based information on the healing components as well as practical, home-tested techniques for using coconut ingredients in everyday dishes. In fact, every recipe in *Coconut 24/7* features at least one coconut ingredient and relies exclusively on organic virgin unrefined (or raw) coconut oil as a healthy fat for sautéing, stir-frying and baking.

While working on this book, I often found myself transported to faraway coastal lands and sun-warmed archipelagos where coconut palms lean into the salt spray and cast their windmill shadows on bleached sand. Their very names—Cook Islands, Tonga, Fiji, Samoa, Solomon Islands, Micronesia, Palau, Thailand,

> Again the cliff yawned, but now with a deeper entry; and the Casco, hauling her wind, began to slide into the bay of Anaho. The cocoa-palm, that giraffe of vegetables, so graceful, so ungainly, to the European eye so foreign, was to be seen crowding on the beach, and climbing and fringing the steep sides of mountains.
>
> ... It was longer ere we spied the native village, standing (in the universal fashion) close upon a curve of beach, close under a grove of palms; the sea in front growling and whitening on a concave arc of reef. For the cocoa-tree and the island man are both lovers and neighbours of the surf. "The coral waxes, the palm grows, but man departs," says the sad Tahitian proverb; but they are all three, so long as they endure, co-haunters of the beach.
>
> **—Robert Louis Stevenson, *In the South Seas* (1896)**

Vanuatu and dozens more—gave inspiration for the tastes that this book enjoys. Every day while testing, I basked in the teasing aroma of just-opened coconut as it met the tang and punch of pineapple or mango or the alluring essence of vanilla. I played with tropical ingredients and cross-pollinated the flavours of Polynesia, Southeast Asia and Oceania with the sensibilities of North American staples.

The recipes here are very different from those found in my previous books. These are infused with the buoyancy and leisurely pace of light-filled days and the warm glow of sunsets over tropical waters. They are an adventurous fusion of exotic and fresh, seasonal and sensible ingredients, and at the same time they offer lots of room for substitutions and experimentation. Here you will find tart apples, pears and other tree fruits tossed with papaya, mango and star anise, all with just the right spike of orange and lemon. Vegetables of every colour brighten the pot and the plate; dips and sauces dance on the tongue; healthy portions of organic meat, chicken, fish and seafood are brought together in unusual combinations that explode on the page and ignite the senses.

As I explored the exceptional health benefits of organic virgin raw coconut oil, I learned of its incredible healing benefits to the skin. Inspired, I simplified my beauty routine (and saved hundreds of dollars!) by using coconut oil as a hair conditioner and to make my own skin and body creams. You'll find these simple spa recipes in Chapter 9.

After an intense immersion in the gentle, renewing and healing ways of coconut, my own relationship with this incredible plant—its water, its meat and its oils—has developed into a full-on love affair. I have experienced first-hand the amazing effects it has on the body—both inside and out. With this book and the products of one plant known and revered by ancient tropical peoples, you, too, can begin an odyssey of delight and delicious adventure.

—Pat Crocker

1

Coconut: The Essentials

The Tree of Life

The coconut tree (*Cocos nucifera*) has been entrenched in the Asia-Pacific culture, diet and traditional healing arts for centuries. In fact, Ayurvedic medicine has been using the oil from "The Tree of Life" as a healing ingredient for some 4,000 years.

In tropical regions where the coconut palm grows in abundance, coconut is a nutritious staple in the diet, rich in fibre, vitamins, minerals and phytonutrients. Folk medicine practitioners have used it for generations to treat a wide range of health conditions, including the following:

- skin conditions (infections, minor wounds, dry skin, bruises, sun protection, abscesses, jaundice, rash, scabies, burns)
- scalp and hair problems (baldness, lice) and as a conditioner
- internal disorders and diseases (asthma, tuberculosis, toothache, dysentery, constipation, earache, kidney stones, gingivitis, inflammation), cough and flu (sore throat, bronchitis, fever, nausea, upset stomach)

Now coconut oil is becoming an important "new" healthy food in many Western and European countries. Modern science has confirmed that it possesses incredible healing properties not present in any other edible oil (see pages 17 to 18). Despite its vilification in the middle of the twentieth century (due to pressure from vegetable oil producers and a lack of understanding about the nature of its fatty acids), the coconut—and, in particular, coconut oil—is beginning to be understood as a "functional food" because it offers several specific health benefits that go far beyond its nutritional constituents.

In the hour of the dusk, when the fire blazes, and the scent of the cooked breadfruit fills the air, and perhaps the lamp glints already between the pillars and the house, you shall behold them silently assemble to this meal, men, women, and children; and the dogs and pigs frisk together up the terrace stairway, switching rival tails. The strangers from the ship were soon equally welcome: welcome to dip their fingers in the wooden dish, to drink cocoanuts, to share the circulating pipe.

—**Robert Louis Stevenson, *In the South Seas* (1896)**

Coconut Water

People who live among coconut palm trees harvest young green coconuts for the exceptionally pure, clear liquid or juice inside the immature nut. The mildly sweet water, called coconut water, is found in abundance in young green coconuts and in much smaller quantities in mature brown coconuts. Young green coconuts can produce between 1-1/2 to 2+ cups of coconut water, whereas a mature coconut may only yield 1/2 cup. Coconut water is a natural component of the growing coconut and is completely different from coconut milk and cream, which are made by pressing shredded coconut meat with plain water (see pages 22 to 23).

Tapping young coconuts for their fresh water delivers more nutrients than the pasteurized (heated) coconut water now available in bottles, cans or Tetra Paks.

Primary Use

Coconut water is consumed as a sports drink or refreshing beverage. Soft, immature coconut meat is often mixed with coconut water to produce a thicker drink or smoothie.

Health Benefits

- Hydrating—lubricates and cushions joints, flushes toxins and waste, and protects sensitive tissues.
- Rich in fibre—fibre isn't easily digested and stays with you longer so it helps you feel full.
- Keeps you regular—easily digested due to medium-chain fatty acids; doesn't require pancreatic enzymes so it puts less of a burden on your digestive system.
- Supports antiviral, antibacterial and anti-inflammatory functions—fatty acids assist these processes, which are crucial to lowering your risk of disease.
- Heart-healthy—contains medium-chain fatty acids (dairy and red meat contain long-chain fatty acids, which increase risk for heart disease).
- Promotes the absorption of other nutrients found in the food you eat.
- Promotes digestive healing and helps prevent chronic inflammation, gas and diarrhea.

- Assists in weight loss—triglycerides are sent directly to the liver for energy so they burn fast and act as a wick to help burn other fats.
- Electrolyte boost—thought to contain the electrolyte equivalent of human blood plasma; vitamins, minerals and enzymes help balance the body's metabolic processes.
- Anti-aging tonic—helps boost immunity and prevent cell damage through oxidation.
- One of the best dietary sources of cytokinins (molecules that protect the cells during healthy cell division and thus help to fight against cancer).

Nutrients in Water from 1 Medium Fresh Young Coconut

3 g saturated fat, 50 mg sodium, 2 g fibre, 15 g sugar, 2 g protein, 17% calcium, 1% iron, with potassium and magnesium

Forms Available

- Whole young green coconuts are sometimes available in specialty food markets and online. These are tapped for the water.
- Husked young green coconuts are available in specialty food markets, in some supermarkets and online. These are tapped for both the water and the immature flesh (see page 9).
- Bottles or Tetra Paks of pure coconut water or coconut water with additives such as sugar and phytochemicals are available at most supermarkets.

Selection and Storage of Fresh Young Green Coconuts

- Look for coconuts that are heavy for their size. Select those that have no sound of water sloshing around inside when shaken because this indicates that the shell is completely full of water.
- Store in the refrigerator or a cool place. Whole young green coconuts will keep in the refrigerator for up to 3 weeks. After tapping a fresh coconut, consume the water immediately or store in the refrigerator for 1 day only to retain the maximum healing nutrients.

Selection and Storage of Coconut Water in Bottles or Tetra Paks

- The product closest to fresh coconut water is raw, unpasteurized coconut water that has been bottled using pressure (not heat) to deactivate bacteria. This product must be refrigerated. Drink or cook with the water immediately after opening or store in the refrigerator for 1 day only after opening.
- Heat-treated water sold in Tetra Paks may be stored in a cupboard and chilled before consuming. Drink or cook with the water immediately after opening or store in the refrigerator for 1 day only after opening.

Health Notes

- All non-refrigerated canned or Tetra Pak liquids—coconut water included—have been sterilized by steam retort, a process that cooks the ingredients and damages much of the healing nutrients.

- Not all canned or shelf-stable Tetra Pak products contain pure coconut water. Always check the label.

Cooking with Coconut Water

- Use as the liquid in smoothies.
- Use in place of plain water or vegetable or chicken broth in soups, stews and other dishes cooked by moist heat (such as poached fruit, fish or chicken).
- Use in place of plain water or broth when cooking rice, vegetables or pasta.
- Substitute for the milk or liquid in baked goods such as bread, cookies and cakes.
- Use in glazes, icing, marinades and sauces.
- Ferment by adding kefir starter to make a probiotic, reduced-sugar drink that helps boost energy, aids digestion and improves immunity.

Caution

Fresh coconut water is clear and the soft, immature flesh inside the shell is white. Do not drink the water or eat the meat from a young coconut that is even slightly pink—this is a sign that the sterile environment inside the coconut has been compromised. Pink water (and meat) from a coconut may have fermented and/or be fizzy and sour-tasting. However, if you purchase *bottled* organic raw coconut water that has turned pink, it is safe to drink. Initially sterile, it has coloured naturally in the bottle due to the antioxidants in the coconut water.

Coconut Meat

Coconuts can grow to a size much larger than a football and are comprised of three distinct layers: The edible meaty or fleshy part (the endocarp) develops inside a hard, brown seed shell somewhat larger than a softball (the mesocarp). To get at the tasty and nutritious seed contents (clear, thin coconut water and thick, firm, white flesh), the fibrous, woody outer brown husk (the exocarp) must first be removed (see facing page).

Raw, fresh coconut meat is white, firm, moist or juicy, and crunchy, with a sweet aroma and taste. It has a pleasant mouth feel due to the presence of fatty acids. It is delicious on its own as a snack or as an ingredient in all kinds of savoury and sweet dishes and beverages. One large coconut yields 1 pound (500 g) of coconut meat, which yields 5 cups of grated coconut.

For many people who live in areas where coconuts grow, fresh coconut meat from mature coconuts is a significant part of their diet. In non-tropical regions, however, fresh coconut can be difficult to find, so it is dried or processed into oil, milk, cream and flour.

When dried, coconut meat is called copra. Before drying and processing, it may be soaked in sugar solutions or corn syrup to sweeten it. It may also have chemical preservatives added, so check the label. Naturally dried coconut is usually unsweetened.

How to Extract Fresh Water from Young Coconuts

1. Using a French knife, puncture the soft shell in a circle around the top of the coconut.
2. Lift off the top of the coconut.
3. Pour liquid into a glass or blender.
4. Cut the coconut in half.
5. Scoop out soft, immature flesh and save it to use in smoothies or cooked cereal. You may keep the flesh for up to 2 days in a refrigerator.

In order to dry fresh coconut meat, it is first sliced, chopped, chipped, flaked, desiccated, ground or shredded, then dried by smoke, the sun or in kilns. If the drying process is not clean, the copra must be purified (refined, bleached and deodorized using high heat) before it can be pressed for oil or sold as an ingredient.

Dried coconut has had all but about 2% of the moisture removed, but it contains almost the same amounts of fat, fibre, protein and other nutrients found in fresh meat (except that, measure for measure, dried coconut is more concentrated than fresh coconut).

Primary Use

Coconut meat is processed for oil (see pages 15 to 16) or used, fresh or dried, in cooking.

Health Benefits

- Dairy- and gluten-free as well as suitable for vegetarian and vegan diets. Although some people may be allergic to coconut, many who suffer from nut allergies find that it is safe to consume.
- Contains coconut oil—both fresh and dried coconut meat consists of roughly one-quarter coconut oil, which has an amazing number of health benefits (see pages 17 to 18).
- Rich in fibre—one cup of fresh coconut meat contains 7.2 g of fibre (29% Daily Value). One ounce of unsweetened desiccated coconut contains 5 g of fibre (18% Daily Value). Coconut (both fresh and dried) contains lower amounts of digestible carbohydrates (starch and sugar) and higher amounts of fibre compared with most other fruits and vegetables. Fibre aids digestion and makes you feel full; it also helps protect against heart disease by reducing cholesterol.
- A good source of iron—one cup of fresh coconut meat contains 11% Daily Value of iron. One ounce of unsweetened desiccated coconut contains 11% Daily Value. Iron is essential for healthy cells, especially red blood cells.
- A good source of potassium—potassium is essential for proper fluid balance in cells as well as proper heart function and muscle growth.
- A good source of copper—copper is important for the production of red blood cells.
- A good source of manganese—manganese helps you to metabolize fat and protein and other nutrients such as iron; it also supports the immune and nervous systems.
- When young coconuts are tapped for the nutrient-rich water, the meat is immature and, as a result, it is very thin, soft and gel-like. This young white flesh is a nutritious, easily digestible food for babies, young children and people recovering from surgery or other health issues. It may also be used in blended smoothies, puddings and other baked desserts, but is not thick or firm enough to grate or chop.

Nutrients in 80 g Dried Coconut

24 g saturated fat, 7 g fibre, 16 mg sodium, 5 g sugars, 3 g protein, iron, vitamins B1, B2, B3, B6, C and E, folic acid, calcium, potassium, copper, manganese, selenium, phosphorus, zinc

Forms Available

- Fresh whole coconut in the husk is available in some specialty stores and online.
- Fresh whole coconut husked but with the hard, brown shell intact is available in many specialty stores, supermarkets and online.
- Fresh flaked or shredded coconut is only widely available where coconuts grow. Grated fresh coconut is often frozen and transported to other non-tropical places.

Selection and Storage

- Select whole coconuts (in the husk or husked and in the shell) that are heavy for their size. Unlike young green coconuts, you should be able to hear the water slosh around the inside of a mature coconut when shaken. Coconuts that have no water inside are old and dry and should be avoided. Avoid whole coconuts with cracked husks or wet/mouldy "eyes."
- Fresh coconuts will keep for up to 4 weeks, but there is no way of telling exactly how long a coconut has been stored up to the point of purchase, so a good rule of thumb is to shake and test for water (see above) and use fresh coconut from a store within 3 days of purchasing it.
- Once a coconut has been cracked and the fresh meat removed, the whole pieces of meat will keep in an airtight container in the refrigerator for up to 5 days. Chopped or shredded fresh coconut will keep for up to 2 days. Fresh coconut may be frozen for up to 3 months if sealed in an airtight container.
- Fresh whole coconuts are available in specialty stores year-round from Florida, Honduras, the Dominican Republic and Puerto Rico, peaking from September through April.

Cooking with Fresh Coconut

- For cooking, fresh coconut meat is either sliced into very thin, long strips or, more commonly, shredded, grated, chipped or chopped. Use a cheese grater for thin, wide shards; a micro-grater for thin, short pieces; or a food processor or electric chopper for either large or small pieces or to produce a coconut powder.
- Use fresh sliced or shredded coconut meat, which is naturally sweet, in a wide variety of recipes, such as appetizers, soups, stews, vegetable dishes, main dishes, salads, marinades, sauces, beverages and desserts.
- Chop or dice fresh coconut for a snack or shred it to use as a garnish for both sweet and savoury dishes.

How to Extract Fresh Coconut from the Shell

1. Using an ice pick or a sharp knife, pierce one or two of the eyes found at the base of the hard, brown shell. Tip the coconut over and drain the water into a bowl or jug. Enjoy immediately as a beverage or transfer to an airtight container and refrigerate for up to 1 day. (See list of uses on page 8.)
2. Freeze the whole coconut for an hour or bake in a preheated 350°F oven for 20 to 30 minutes. (Freezing or heating the coconut causes the paper-thin, brown skin on the inside of the meat to shrink away from the hard seed shell, making is easier to peel the meat.)
3. If using the oven method, set the coconut aside to cool.
4. Wrap the coconut in a dishtowel and, using the flat edge of a hammer, tap all over to crack the shell open. Using a metal spatula or blunt-tipped knife, pry the meat from the inside of the broken shell pieces.

Caution

There have been cases of recall of freshly grated coconut due to bacterial contamination.

Dried Coconut

Forms Available

- Sweetened or unsweetened coconut: Ground, desiccated (also called "fine"), fine macaroon, grated, shredded, flaked, chipped, diced or toasted—various forms are widely available in grocery stores and online.
- Creamed coconut: Dried unsweetened coconut is ground to a semi-solid paste and pressed into a block. Depending on the brand, it may be smooth or somewhat gritty. It imparts a more concentrated coconut flavour than dried coconut in any other form. Don't confuse creamed coconut (sometimes called cream of coconut) with coconut cream (see page 20) or with powdered coconut milk or cream. Creamed coconut is packaged in a box and usually found in the baking ingredients section of the supermarket.

Forms of Dried Coconut

1. Large flakes
2. Diced
3. Dessicated or fine
4. Shredded
5. Grated
6. Toasted

Dried Coconut

Selection and Storage of Dried Coconut

- Dry your own coconut using a dehydrator (follow the manufacturer's instructions) or source naturally dried unsweetened organic coconut for use in most of the recipes in this book. Most whole food or health food stores carry high-quality dried coconut.
- Avoid dried coconut that has added preservatives.
- Dried coconut will keep for up to 3 months if kept in an airtight container stored in a cupboard (because the oil is shelf-stable).

Cooking with Dried Coconut

- Use naturally dried shredded coconut as you would fresh coconut in a wide variety of appetizers, soups, stews, vegetable or curry dishes, main dishes, salads, marinades, sauces, beverages and desserts.
- Unsweetened coconut, which is mildly sweet, is best in savoury recipes for curries, chutneys, soups, stews, tagines and other meat dishes.
- Sweetened dried coconut lends a light texture to baked goods due to its higher moisture content from the added syrup.

Health Notes

Dried coconut may be soaked in sugar solutions or corn syrup to sweeten it, and nitrites or sulphites are sometimes used as preservatives. Always check the label.

Creamed Coconut

Selection and Storage of Creamed Coconut

- Look for organic unsweetened creamed coconut.
- Store at room temperature. Creamed coconut will keep for a year or more if unopened. Store opened creamed coconut in an airtight container in the refrigerator for 1 year or longer.

Cooking with Creamed Coconut

- Use to make an easy coconut cooking liquid: Mix 2 tablespoons creamed coconut with 1 cup coconut water or plain water. Add the mixture to soups, stews, sauces, dips or any recipe that calls for vegetable broth or coconut milk.
- Stir 1 tablespoon into smoothies, curries, stews, gravy, soups, dips and sauces.
- Use in icing and as a filling for cakes, breads, muffins, cookies and chocolates to impart a coconut flavour.
- Use to make coconut ice, yogurt and ice cream.

Health Note

Sometimes sugar and other additives are added to creamed coconut. Always check the label.

Coconut Oil

Coconut oil is the essential constituent in both coconut water and coconut meat

that gives them their healing properties. Until recently, coconut oil, which is a saturated fat (solid at room temperature), was considered a poor dietary choice. We now know that coconut oil is an extremely important and healthy fat to include in our daily diets. Do not confuse saturated coconut oil, which is composed of short- and medium-chain fatty acids extracted from a plant source, with saturated fats from hydrogenated oils or animal sources, which cause long-term damage to the body.

Our bodies actually need high-quality polyunsaturated and saturated fats that are high in short- or medium-chain fatty acids to keep our serum cholesterol low. The problem with mono- or polyunsaturated oils such as olive or canola oil is that they are unstable and susceptible to oxidation and free radical formation unless they are unrefined, cold-pressed, stored properly (in transportation, at the store and in the kitchen) and used for raw or low-heat recipes. Coconut oil is so high in lauric acid that it is incredibly stable and it does not oxidize or break down during transportation, storage or at moderately high cooking temperatures. Research by nutritionist and biochemist Mary G. Enig, Ph.D., suggests that lauric acid and other complex ingredients in coconut oil actually reduce LDL (bad) cholesterol and improve HDL (good) cholesterol.

Methods of Processing

Two things determine the ultimate nutrient level of any oil: the quality of the plant source and the method of extracting the oil from it. Coconuts are a sustainable crop, and they do not need fertilizers or chemicals to grow. The "organic" designation for coconut products is more important when it is used to describe the way that the coconuts are processed (if chemicals have been used to separate the oils from the solids, the resulting product—either copra or the oils—is not organic). Unfortunately at the present time, "organic" only refers to growing methods, not the oil production process.

"Virgin" is the term used to describe oils that have been pressed in a machine that is cold—they are the first oils to be produced before the press heats up. Virgin oils may be expeller pressed and then further processed either by heating or fermenting to separate the water out, so it pays to research a brand before buying.

Oil that is extracted from copra may be purified (refined, bleached and deodorized) using high heat, and sometimes sodium hydroxide (lye) is added to prolong shelf life. These lower-quality (and usually lower-priced) refined coconut oils are, therefore, devoid of most, if not all, health benefits.

On the other hand, oil that is extracted from fresh coconuts by mechanical or expeller pressing or by centrifuge does not require chemical refining (bleaching or deodorizing). If heat has not been used, these methods do not alter the natural components of the oil in any way and they retain all of the health benefits (see pages 17 to 18).

The centrifuge method of separating the oil from the water (the healthiest because no heat is used) is a two-step process. First, fresh grated meat is pressed to produce coconut milk. The milk is then spun to separate the water and solid matter from the oil. No heat, chemicals or deodorizers are used in this process, and it is often referred to as "extra-virgin."

Extra-virgin or virgin raw unrefined oil from fresh organic coconuts is the very best you can buy. (It is comparable in price to other high-quality organic virgin raw oils.) It retains the fatty acids, nutrients and phytonutrients and delivers the health benefits you want. It is stable (won't break down by oxidation) at room temperature and is safe to use in cooking at moderate temperatures (350°F). Because it is stable, you don't have to worry about how it was transported or stored.

Primary Use

Coconut oil is primarily used in cooking, but it is also an ingredient in skin and hair products.

About Hydrogenated Fats

When any oil is broken down by heat or chemicals or when it has oxidized (the oxygen molecules have been ravaged by free radicals), its health-giving fatty acids are transformed into trans fatty acids, a completely different and heart-damaging substance. This process is known as hydrogenation: the structure of the oil is altered from a polyunsaturated form to a saturated one by the addition of hydrogen. Hydrogenation is used to make polyunsaturated fats solid at room temperature and to stabilize them. Because they are carcinogenic and can cause heart disease and other major problems, all hydrogenated fats, including coconut oil, that have undergone this process should be eliminated from your diet.

Health Benefits of Organic Virgin Raw Coconut Oil

- Helps you lose weight—its medium-chain fatty acids help regulate thyroid function, balance blood sugar levels, increase energy and stimulate metabolism, all of which encourage weight loss.
- Lowers cholesterol—almost half (49%) of the fatty acids in coconut oil are lauric acid (converted to monolaurin in the body), which helps lower LDL (bad) cholesterol and increase HDL (good) cholesterol.
- Reduces your risk of heart disease—medium-chain fatty acids are rapidly absorbed almost directly into the liver and are converted into energy instead of being stored either in adipose (fat) tissues or deposited in the arteries (as are the long-chain fatty acids in animal fats).
- Stable at room and moderate cooking temperatures—it won't oxidize and form dangerous free radicals (as do unsaturated fats).
- Suitable for those with diabetes—the medium-chain fatty acids in coconut oil do not raise blood sugar levels and they keep levels even by providing a steady source of energy without stimulating the release of insulin. Steady blood sugar levels help prevent hypoglycemia, a precursor to type 2 diabetes.
- Supports your liver—the liver is the clearing house for the body's toxins, hormones, bile and other substances. Coconut oil's antimicrobial fatty acids cleanse the liver and help protect it from damage from free radicals.
- Boosts your daily energy—the medium-chain fatty acids pass directly to the liver, where they are burned for energy; 2 teaspoons of coconut oil a day will supply you with steady energy.
- Improves nutrient absorption—studies have shown that coconut oil enhances the body's ability to absorb vitamins, minerals, antioxidants and other phytonutrients from food, especially fats and fat-soluble vitamins. These studies suggest that food cooked in coconut oil will help transport nutrients from the food through the stomach lining into the bloodstream. Consuming coconut oil on its own or in food helps prevent vitamin and mineral deficiencies.
- Improves digestion—because medium-chain fatty acids in coconut oil are smaller than the long-chain fats found in other oils and animal fat, they require less energy and fewer enzymes to break down for digestion. This puts less strain on all the digestive organs, especially the pancreas and gall bladder.
- Decreases inflammation—inflammation is a major factor contributing to the development of disease. According to a study from the Institute of Human Nutrition at the University of Southampton, UK, coconut oil may be useful in the treatment of both acute and chronic inflammation and inflammatory diseases, including heart disease.
- Improves your immune system—organic virgin raw coconut oil is high in lauric, capric and caprylic acids. These fatty acids are antimicrobial, antifungal and

antiviral, meaning they combat the minute organisms that cause disease. Lauric and capric acids in coconut oil help treat intestinal yeast infections such as candidiasis. Lauric acid is converted in the body to monolaurin, which has been shown to inactivate or protect against viruses such as HIV, measles, herpes simplex virus, flu and rubella.

- Helps prevent breast, colon and other cancers—coconut oil helps to keep the blood void of free radical damage, which contributes to degenerative diseases such as heart disease and cancer. Oxidation occurs when hydrogenated oils (shortening and margarine) or animal fats (which are long-chain fatty acids) are consumed, so eating coconut oil not only offers protection against free radicals, it replaces other unhealthy fats.
- Improves Crohn's disease, irritable bowel syndrome, ulcerative colitis, gall bladder disorders, pancreatic disorders and other digestive disorders—the antiviral, antimicrobial and antibacterial constituents in coconut oil play a role in neutralizing bacterial infection, and the anti-inflammatory properties soothe inflammation in the digestive tract.
- Helps prevent osteoporosis and supports the development of strong bones and teeth—the antioxidants found in coconut oil protect the bones from destruction by oxidizing free radicals.
- Moisturizes, heals and protects the skin—it absorbs quickly into the skin, does not feel greasy or sticky, and has been found to diminish and prevent wrinkles, moisturize and alleviate minor burns, rashes and insect bites.
- Conditions your hair and helps control dandruff—using coconut oil either before or after shampooing your hair can leave it soft, shiny and full.

Caution

Because the medium-chain fatty acids in coconut oil require few or no enzymes to break them down in the intestines, coconut oil can cause mild forms of diarrhea in some people. It is wise to start with small amounts when cooking with or taking coconut oil as a supplement for the first time.

Nutrients in 100 g Coconut Oil

86 g saturated fat (short- and medium-chain saturated fatty acids including lauric acid), 1.8 g polyunsaturated fat, 6 g monounsaturated fat

Forms Available

Coconut oil is available in pint or quart glass jars or in larger quantities in plastic tubs.

Selection and Storage

- Use only certified organic, unrefined or raw coconut oil. If you can determine the method of extraction, choose the oil that has been cold-press expelled and centrifuged to remove the water.
- Two indicators of good-quality coconut oil are a higher price and a pleasing coconut scent. It is expensive to extract the

oil in a way that is healthy. Oils that have been refined, bleached and deodorized do not carry the distinctive smell of coconut and do not contain the healing properties of organic virgin raw coconut oil.

- Store coconut oil in a cool cupboard and it will keep for up to 3 years. Its stability makes coconut oil resistant to oxidation and rancidity. For convenience, transfer 1 to 2 cups from the original container to a glass container with a tight-fitting lid and place it close to the stove. This will ensure that you have a constant supply of softened coconut oil at hand for cooking.

Cooking with Coconut Oil

- In addition to its many health benefits, virgin raw coconut oil is heat stable at moderate temperatures—its smoke point is 350°F, the highest of any unrefined vegetable fat. This makes it the perfect choice for almost all cooking methods, including stir-frying, baking, roasting and pan-frying. Most other oils, including olive oil, are not stable at moderate temperatures, especially when used to fry food, and they can break down into highly carcinogenic trans fatty acids.
- Keep the health benefits of coconut oil by cooking over medium to low temperatures on the stovetop and limiting oven temperature to 350°.
- Coconut oil is a healthy all-purpose oil—replace the cooking oils you currently use with high-quality organic virgin raw coconut oil. Keep a small amount of high-quality organic virgin raw polyunsaturated oil, such as walnut, olive or sunflower oil, for use in salad dressings or other raw dishes when you want an oil that imparts a different flavour and one that won't firm up at cooler temperatures. Always store polyunsaturated oils in the refrigerator.
- Coconut oil is solid at room temperature but easily melts to oil at 76°F.
- Use as a butter replacement: Drizzle on vegetables or spread over bread (or whip half butter and half softened coconut oil).
- Stir 1 tablespoon into hot or warm drinks such as cocoa, tea, coffee, chai and cider.
- Add 1 tablespoon to smoothies.
- Stir 1 teaspoon into cooked rice, amaranth, quinoa, oatmeal and other grains.
- Use to grease pans for baking.

How to Soften Solidified Coconut Oil
Bring a small saucepan of water (2 to 3 inches) to boil over high heat. Remove the lid from the jar of coconut oil and stand the jar in the water. Within a few seconds, the oil will be soft enough to measure. Recipes typically call for two types of coconut oil: softened and melted.

Softened coconut oil—The oil has softened to a consistency that is similar to butter at room temperature. You can scoop softened coconut oil with a spoon.

Melted coconut oil—The oil has melted to a clear liquid similar to corn oil or olive oil. You can pour melted coconut oil into a liquid measuring cup or spoon.

Caution

Do not heat coconut oil above 350°F or past its smoke point—excessive heat will cause it to break down into oxidizing trans fatty acids and destroy its health benefits.

Coconut Milk and Coconut Cream

Coconut milk and coconut cream are both white, creamy liquids with a nutty-coconut flavour. You can make them at home by mixing water with fresh or dried coconut meat and then squeezing it to extract the thick, rich milk or cream (see pages 22 to 23). Coconut cream is the result of the first pressing—it is thicker, with more oil and solids and less water. As more water is added to extract solids from the meat, the product gets thinner—this is coconut milk.

Commercially produced milk and cream are pressed from copra, but gums are added for extra thickness and texture. Some also have added preservatives; check the labels.

If commercially canned coconut milk is refrigerated or if it has not been disturbed for some time, the lighter, fat-rich cream floats to the top, especially in brands that do not use emulsifiers or thickeners. If you have a use for the cream, scoop it out of the can before using the milk in recipes; if not, shake the can before opening and the fat will be distributed evenly into the milk.

sweetened and contain added guar gum, xanthan gum, sodium carboxymethyl cellulose, polysorbate 60 and sulphites.

Do not confuse the coconut beverage sold in Tetra Paks with coconut milk. It is sweetened with cane juice and some brands have most or all of the following added: inulin, carrageenan, gellan gum, sea salt, xanthan gum and guar gum. In addition, there is little coconut oil in these products (about 5 g of saturated fat per cup). As with all liquids packaged in Tetra Paks, the contents have been heated to a high temperature to kill micro-organisms, thus destroying any of the natural benefits in the coconut oil.

Primary Use of Coconut Milk and Coconut Cream

Coconut milk and cream are primarily used in cooking.

Health Benefits of Coconut Milk and Coconut Cream

Drinking—or cooking with—coconut milk and cream will deliver the benefits of coconut oil to the extent that it is in the product. Because coconut cream is usually higher in oils, it will be slightly more beneficial than coconut milk. "Lite" or fat-reduced coconut milk will be significantly lower in the health-giving oil.

Caution

Canned coconut milk and cream may contain bisphenol A. They may also be

Forms Available

- Canned coconut milk and cream are widely available. Choose organic coconut milk and coconut cream—they offer the best-quality fat and creamier liquids.
- The coconut milk beverage available in a Tetra Pak contains 5 g of saturated fat per cup, is sweetened with cane juice and may contain inulin, carrageenan, gellan gum, sea salt, xanthan gum and guar gum. Because of the additives and lower fat, I do not recommend using this product in cooking.
- Avoid "lite" coconut milk—it has had roughly 60% of the nourishing coconut oil removed.

Selection and Storage

- Making your own coconut milk and cream ensures they will be organic (if an organic coconut is used) and free of gums and chemical additives. If you must use canned coconut milk, look for organic products packaged in BPA-free cans and always check the label for unwanted ingredients.

- Cans and Tetra Paks may be stored in a cupboard until ready to use.
- Refrigerate freshly made coconut milk and cream—and the milk or cream from opened cans and cartons—in an airtight container for up to 3 days.
- Fresh coconut milk and cream may be frozen and will keep for up to 3 months in the freezer but may separate. Once thawed, shake the container to distribute the fat evenly into the liquid; use immediately or refrigerate for up to 3 days.

Cooking with Coconut Milk

- Use coconut milk in curry and Thai dishes.
- Use to replace water or vegetable or chicken broth in soup, stew and other moist-heat cooking methods.
- Substitute for dairy milk in smoothies, baked goods, sauces and dips.
- Use in custards and other desserts.

Cooking with Coconut Cream

Coconut cream is thicker than coconut milk and is an excellent replacement for dairy cream in sauces, dips and desserts.

How to Make Coconut Cream and Milk

The first press will yield a thick, rich cream. The second press will produce a thinner milk. Experiment with quantities to determine the perfect consistency for your needs.

To make cream from fresh coconut: For every 3 cups shredded fresh coconut meat, use 2 cups hot water (or scalded dairy milk).

Method A: Line a bowl with cheesecloth, add coconut and pour water over. Let steep for 30 minutes. Gather the cheesecloth up and squeeze the liquid through it, collecting the cream in the bowl. Reserve the coconut solids to make milk.

Method B: In a blender, combine 3 cups shredded coconut and 2 cups hot water. Process for 1 to 2 minutes or until coconut is finely chopped. Line a sieve with cheesecloth and place over a bowl. Pour coconut mixture into sieve and, using the back of a wooden spoon, press out the cream. Reserve the coconut solids to make milk.

Method C: Use a heavy-duty juicer to separate the juice from the pulp. The result is a smaller amount of rich, thick coconut cream. If desired, dilute this with either coconut water or plain water to make coconut milk. Use the pulp in soups, stews and curry dishes or to make face or body scrubs.

To make milk from fresh coconut after making cream: After making coconut cream following any of the methods outlined, you can use the reserved coconut solids to make milk. In a saucepan, combine the reserved coconut with 2 cups of water and bring to a boil over high heat. Reduce the heat and simmer for 5 minutes. Set aside to cool. Transfer the cooled mixture to a blender and blend for 1 minute. Line a sieve with cheesecloth and place over a bowl. Pour coconut mixture into the sieve and, using the back of a wooden spoon, press out the milk.

To make milk without first making cream: For every 2 cups shredded fresh coconut meat, use 4 cups hot water (or scalded milk) and follow directions for methods A or B on facing page.

To make cream from dried coconut: For every 2 cups of shredded dried coconut meat, use 1 cup hot water (or scalded milk).
Method A: Line a bowl with cheesecloth, add coconut and pour water over. Let steep for at least 30 minutes or up to 4 hours. Gather the cheesecloth up and squeeze the liquid through it, collecting the cream in the bowl. Reserve the coconut solids to make milk.
Method B: In a blender, combine coconut and water. Process for 1 to 2 minutes or until finely chopped. Line a sieve with cheesecloth and place over a bowl. Pour coconut mixture into sieve and, using the back of a spoon, press out the cream. Reserve the coconut solids to make milk.

To make milk from dried coconut: After making coconut cream following any of the methods outlined, you can use the reserved coconut solids to make milk. Follow the directions for making milk from fresh coconut (on facing page).

To make milk without first making cream: For every 2 cups shredded dried coconut meat, use 2 cups hot water (or scalded milk) and follow directions for methods A or B at left.

To make milk from creamed coconut: In a blender, combine 2 parts water and 1 part creamed coconut and blend until coconut is evenly dispersed in the water. Use as is (with gritty coconut). If you prefer a smoother milk, line a bowl with cheesecloth and pour the blended mixture into it. Gather the cheesecloth up and squeeze the liquid through it, collecting the milk in the bowl. Discard the fine coconut grounds.

Notes

- For a thicker cream, refrigerate freshly made (or canned) cream and then skim the soft, thick, fatty layer that rises to the top (if canned, flip the can upside down, open and scoop the solids out). Use the remaining liquid as milk.
- Use fresh coconut milk or cream immediately or refrigerate in an airtight container for up to 3 days.
- Instead of discarding the coconut that has been used to make coconut milk or cream, freeze it and use it in curries, soups and stews or to make face and body scrubs (see Chapter 9).

Coconut Flour

Coconut flour is dried, defatted and finely ground unbleached coconut meat. It does not require the addition of preservatives, so check the label to make sure the brand you buy is pure coconut flour. Coconut flour is suitable for use in recipes calling for flour and yet it is gluten-free. Like the oil, it is very stable and should retain its nutrient profile for the time you will be storing it.

Primary Use

Coconut flour is primarily used in cooking.

Health Benefits

- High in fibre (higher than any other flour)—2 tablespoons of coconut flour delivers 5 g of fibre, which helps prevent constipation, absorbs toxins and helps manage diabetes and prevent degenerative diseases such as heart disease and cancer.
- Low in digestible carbohydrates (glucose)—2 tablespoons of coconut flour contains 8 g of carbohydrates (less than most fruit and vegetables).
- Gluten-free.

Caution

Some brands contain sulphites (carcinogenic additives). Always check the label.

Nutrients in 1/4 Cup of Coconut Flour

10 g (40% DV) fibre, 4 g fat, 60 mg sodium, 16 g carbohydrates, 4 g protein, iron (4% DV)

Forms Available

Coconut flour is available fine or coarse, both labelled as flour. It is available in some supermarkets and most bulk, whole food and health food stores.

Selection and Storage

- As with coconut oil, the quality of coconut flour depends on the quality of the coconuts and the method used to mill them. Look for organic, unbleached, raw unsweetened coconut flour.
- Unopened, coconut flour is shelf-stable for up to 1 year stored at room temperature and for up to 2 years if refrigerated

or frozen. Reseal and refrigerate or freeze after opening.

Cooking with Coconut Flour

- Finely milled from dried coconut meat, coconut flour has a wheat-like consistency that makes it suitable for use along with wheat or other kinds of flour in recipes calling for flour. Because it is gluten-free, it's great to add to gluten-free flour for those who suffer wheat allergies or who are following a gluten-free diet. Like the oil, it is very stable and should retain its nutrient profile during storage.
- Its light coconut flavour blends well with wheat or other types of flour in savoury baked goods.
- Coconut flour can replace up to 20% of the flour in cakes, cookies, pies and quick breads, but you will need to add an equal amount of liquid to allow for its hygroscopic (water-absorbing) quality. Note that coconut flour will increase the volume due to its high fibre content.
- Add up to 10% coconut flour to yeast breads but add an extra egg or 1/4 cup milk.
- Because it contains fibre, coconut flour works well as a coating for chicken, fish or seafood—use it in place of regular flour or cornmeal.
- Coconut flour does not have the thickening power of gluten (as does wheat flour), but you can use coconut flour to thicken sauces, gravy and curry dishes because the fibre swells as it absorbs liquid from the mixture. Sauces are slightly gritty and not as smooth as those made with wheat flour.
- Baked goods made with coconut flour are more crumbly than those made with wheat flour.

Coconut Sap Products: Sugar Crystals, Nectar, Vinegar, Aminos

Trees and plants produce a sticky fluid called sap that is transported from the roots toward the leaves (or from where carbohydrates are produced to where they are stored). In addition to the food products derived from the coconut seed, the tree's naturally occurring sap is processed into healthy cooking ingredients. Coconut sugar crystals and nectar are derived entirely from the raw sap, called *tuba*, which is collected from coconut flowers or the stem that feeds the flowers and fruit (coconuts) of the coconut tree.

Fresh coconut sap (directly out of the tree) consists of 0.5% glucose, 1.5% fructose, 16% sucrose and water. It also contains the following nutrients: 17 amino acids, vitamin C, broad-spectrum B vitamins and trace minerals.

Health Benefits

- Coconut sap is low on the glycemic index—the presence of fibre and inulin are the key factors that maintain the glycemic index at an average of 35 in the crystals and nectar.
- Raw sugar crystals and nectar are rich in enzymes, the nutrients that help the body with various digestive functions.
- The fructose in raw coconut nectar is not processed through the liver, unlike corn and agave syrups, where the fructose can be as high as 90% and is converted into lipids as a by-product.

Coconut Sugar Crystals

Having no coconut aroma or flavour, coconut sugar crystals are tiny beads with a texture that is drier than brown sugar and have the pleasing flavour of caramel and molasses.

Primary Use

Coconut sugar crystals are primarily used in cooking.

Nutrients in 1 Tablespoon Coconut Sugar Crystals

7 g carbohydrates (sugars: 8% to 10% glucose, 10% to 12% fructose and nearly 74% sucrose), 10 mg sodium, and a proportion of the 17 amino acids, vitamin C, broad-spectrum B vitamins and trace minerals from the sap

Forms Available

Coconut sugar crystals are only available in a dry, dark brown crystal form. You can find them in some supermarkets and many specialty, bulk, health and whole food stores.

Selection and Storage

- Look for organic, unrefined, unbleached coconut sugar crystals.
- Store in an airtight container at room temperature for up to 6 months.

Cooking with Coconut Sugar Crystals

- You can substitute coconut sugar crystals for granulated white or brown sugar in recipes.
- Coconut sugar crystals can't be caramelized like granulated white sugar because they burn at a much lower temperature than refined cane sugar. (Caramelizing requires a temperature from 320°F to 356°F, which is too high for coconut sugar crystals.)

Coconut Nectar

Thick and rich, coconut nectar is a dark brown, tartly sweet syrup with heavy malt and molasses overtones.

Primary Use

Nectar is primarily used in cooking.

Nutrients in 1 Tablespoon Coconut Nectar

13 g carbohydrates (same sugar ratio as crystals), 20 mg sodium, and a proportion of the 17 amino acids, vitamin C, broad-spectrum B vitamins and trace minerals from the sap

Forms Available

Coconut nectar is available in 12-oz (375 mL) or smaller bottles or plastic vacuum packs. You can find it in some supermarkets and many specialty, health and whole food stores.

Selection and Storage

- Opt for organic, unrefined, unbleached coconut nectar.
- Store in original airtight container (or transfer to an airtight container) at room temperature.
- Coconut nectar will keep for several years if moisture does not penetrate the container.

Cooking with Coconut Nectar

- You can substitute an equal amount of coconut nectar for other liquid sweeteners, such as honey or maple syrup, in all recipes.
- Use to sweeten beverages and teas.
- Drizzle over cereal and fruit.

Coconut Vinegar

Robust, with no coconut aroma or flavour, coconut vinegar is the fermented sap (known as tuba) taken from the stem that feeds the coconut flower and fruit (coconut). As coconut vinegar ages, the taste becomes sharper, and it turns from cloudy white to light yellow to a clear light brown. Sediments and vinegar "mother" will settle on the bottom of the bottle as the vinegar ages, but these do not interfere with its quality or use; simply shake the bottle before measuring into recipes.

Primary Use
Coconut vinegar is primarily used in cooking.

Health Benefits
Raw coconut vinegar is a live food product that delivers the enzymes, amino acids and trace minerals found in the sap.

Nutrients in 1 Tablespoon Coconut Vinegar
5% sodium and a very small proportion of the 17 amino acids, vitamin C, broad-spectrum B vitamins and trace minerals from the sap

Forms Available
Coconut vinegar is usually sold in 12-oz (375 mL) bottles or smaller. You can find it at some specialty or food stores and many whole food and health food stores.

Selection and Storage
- Unfiltered and unheated (raw) coconut vinegar is preferred over pasteurized, which kills all live enzymes.
- Store at room temperature.
- Will keep for a year or longer.

Cooking with Coconut Vinegar
- Substitute for apple cider vinegar in recipes. Use in dressings, marinades, sauces, dips and other sweet-sour recipes that call for vinegar.
- For a refreshing drink, combine 1 to 3 tablespoons of coconut vinegar with 1-1/2 cups warm water and coconut nectar or honey to taste. Chill and add ice if desired.

Coconut Aminos

Amino acids are the building blocks of protein, which are the nutrients needed by the body to build and maintain organs and tissues. Coconut liquid aminos are available as a seasoning sauce made from the fresh sap of the coconut tree, sometimes with a little added mineral-rich sea salt. They are fermented for approximately 8 weeks and aged.

Primary Use
Aminos are primarily used as a flavouring or seasoning in cooking.

Health Benefits
- Raw aminos deliver beneficial enzymes and bacteria that are alive and viable.
- The aging/fermentation process creates naturally occurring beneficial probiotics.
- They contain approximately 65% less sodium than other non-coconut brands of soy-based aminos.

Nutrients in 1 Teaspoon Coconut Aminos
5% sodium and a very small proportion of the 17 amino acids, vitamin C, broad-spectrum B vitamins and trace minerals from the sap

Health Note
While no sulphites are added to most raw coconut aminos, sulphites are a naturally occurring compound (found on grapes, onion, garlic and other plants) that nature uses to prevent microbial growth. As aminos go through a natural aging process, naturally occurring sulphites are

more likely to exist in them, but in trace amounts. This is true of most aged products, such as soy and tamari sauce.

Forms Available

Coconut aminos are usually sold in 8-oz (250 mL) bottles or smaller. You can find aminos in health and whole food stores and online.

Selection and Storage

- Look for raw coconut aminos, which means that beneficial enzymes and bacteria remain alive and viable.
- Store at room temperature. Will keep indefinitely.
- Because raw aminos are a live, aged product, you may come across some bottles that are more effervescent than others due to continued fermentation within the bottle from the natural yeast. If this happens, pressure builds up inside the bottle and, when opened, it can fizz over. To prevent this, place aminos in the refrigerator to settle the contents prior to opening for the first time. You do not need to store opened aminos in the refrigerator.

Cooking with Coconut Aminos

- Coconut aminos are used to add saltiness and flavour to a wide range of recipes. The clear brown liquid has no coconut flavour or aroma and is similar in taste to soy or tamari sauce.
- Substitute for soy or tamari sauce or for other soy aminos.
- Serve with rice or sushi or use in stir-fry dishes, casseroles and other recipes, especially Asian dishes.
- Use in drinks, sauces, marinades, dips, dressings and soups.

2

Breakfast

Banana Coconut Oatmeal

Makes 2 servings

This satisfying breakfast dish may be served either hot or at room temperature with dairy or non-dairy milk. Top with chopped or toasted nuts or seeds; plain or toasted coconut; raisins; dried cranberries or dried chopped apricots; or fresh seasonal berries or chopped tree fruit.

1 can (14 oz/398 mL) coconut milk
1-1/2 cups large-flake rolled oats
2 bananas, thinly sliced
1 tsp softened coconut oil
1 tsp coconut nectar or honey
1/4 tsp ground cinnamon

1. In a saucepan, bring the milk to a boil over high heat. Stir in the oats, bananas, oil, nectar and cinnamon. Reduce heat and simmer, stirring frequently, for about 8 minutes or until all of the liquid has been absorbed.

Rolled oats (*at left*) are oat kernels or "groats" that have been steamed and then flattened into large flakes by heavy rollers. Often called large-flake or old-fashioned oats, they take longer to cook than quick or "instant" rolled oats, which have been steamed longer and rolled thinner. Cooked instant oats are softer and somewhat mushy, whereas large-flake oats hold their shape and have a chewier texture after cooking. You can use instant rolled oats in this recipe, but you will need to follow the package directions and reduce the cooking time significantly.

If you happen to have leftover cooked oatmeal, press it into a resealable container and refrigerate it overnight. The next day, pop the porridge "loaf" out of the container and cut into 1/2-inch slices. Fry the slices in an oiled skillet (I use coconut oil) over medium heat for about 4 minutes on each side or until browned on both sides. Serve with fresh fruit or maple syrup. Thanks to John Haidet and his wife, Kate, for introducing me to these while we were travelling in the Cotswolds, England—oatmeal slices are a delightful and creative way to "waste not."

Black Rice Torte

Makes 6 to 8 servings

Black rice was once called the forbidden rice of China's vast empire because only the emperor and his family held the right to eat it. Given that black rice is loaded with antioxidants, this was indeed a great privilege. Now this amazing purple-black rice is widely available at many supermarkets, Asian grocery stores and whole food markets. This is a weekend brunch or "night before" weekday recipe. Finish to the end of Step 5 and spread mashed potatoes evenly overtop of the torte. Cover and refrigerate until the next morning. Bring to room temperature and bake, following the directions in Step 6.

In this recipe, I have mixed equal amounts of black and short-grain brown rice, but you can use a blend of any of the following: long- and short-grain brown, wild, black Japonica, and red and mahogany rice. Look for gourmet blends in finer supermarkets or whole food markets.

You can use button, white or cremini mushrooms, but I prefer to use a mix of shiitake, cremini and oyster mushrooms for a richer flavour.

1/2 cup black rice
1/2 cup short-grain brown rice, rinsed and drained
2 large potatoes, cut into 1-inch cubes
2 tbsp butter
1/4 cup milk
2 tbsp melted coconut oil
1 onion, coarsely chopped
1 leek, white and light green top, coarsely chopped
1 lb (500 g) mixed mushrooms, coarsely chopped
1 tbsp fresh thyme leaves
1 tsp sea salt
1/2 cup finely chopped walnuts
1 egg, lightly beaten
1 cup shredded cheddar cheese

1. In a saucepan, bring 2 cups salted water to a boil over high heat. Add black and brown rice. Cover, reduce heat and simmer for 40 minutes or until chewy-tender. Remove from heat, uncover and let stand for 3 minutes to allow the rice to absorb any remaining liquid. Fluff with a fork and set aside to cool.
2. In another saucepan, cover potatoes with water and bring to a boil over high heat. Reduce heat and simmer for 12 minutes or until potatoes are fork-tender. Drain and rinse with cold water. Return to saucepan. Add butter and milk and heat over medium heat. Mash using a potato masher. Set aside.
3. Meanwhile, in a skillet, heat oil over medium heat. Sauté onion and leek for 6 minutes or until soft. Add mushrooms, thyme and salt. Sauté for 8 minutes or until mushrooms are soft and liquid has evaporated. Stir in nuts. Set aside.
4. Preheat oven to 400°F. Grease and line a 9-inch springform pan with parchment paper.
5. Combine egg and cooked rice and pat evenly into the bottom of prepared pan. Sprinkle with cheese. Spread mushroom-nut mixture evenly overtop. Bake in preheated oven for 35 minutes.
6. Remove torte from oven and spread mashed potatoes evenly overtop. Bake for 10 minutes or until lightly browned. Let stand for 10 minutes before serving.

Coconut Granola

Makes 6 cups

Give yourself a boost of energy with this satisfying granola packed with nuts, seeds, oils and nectar. The combination of protein, fibre and healthy fat delivers more staying power than most commercial granolas, which aren't as high in protein-dense ingredients. Enjoy a spoonful with yogurt, cooked cereal or fruit or add to muffin mixes and use as a loose topping for tarts and bars.

- 4 cups large-flake rolled oats
- 1/2 cup natural bran or bran flakes cereal
- 2 cups dried shredded or flaked coconut
- 1 cup chopped almonds
- 1/2 cup sunflower seeds
- 1/4 cup sesame seeds
- 1/2 cup softened coconut oil
- 1/3 cup coconut nectar or liquid honey
- 2 tsp ground cinnamon
- 1/2 cup chopped dried apricots
- 1/2 cup raisins
- 1/2 cup dried cherries

1. Preheat oven to 375°F. Lightly oil 2 rimmed baking sheets.
2. On 1 prepared baking sheet, evenly spread oats and bran. On the remaining sheet, evenly spread dried coconut, almonds, sunflower seeds and sesame seeds. Stagger the sheets in preheated oven and toast for 8 minutes. Remove sheet of nuts and seeds from oven and set aside to cool. Stir grains and continue toasting for another 6 to 8 minutes or until lightly browned.
3. Meanwhile, in a small saucepan, heat oil, nectar and cinnamon over medium heat for 3 to 5 minutes or until simmering. Turn off heat and let stand on burner to keep warm.
4. Transfer toasted grains to a large bowl. Stir in the toasted nuts and seeds. Drizzle with warm nectar mixture. Add apricots, raisins and cherries and lightly stir to coat evenly. Set aside to cool. Store granola in an airtight container in the refrigerator for up to 3 months or freeze for up to 6 months.

Blueberry Granola Muffins

Makes 12 muffins

Soft blueberries and crunchy granola are a winning combination in these muffins sweetened with coconut nectar and grated coconut—a nice change from regular muffins, which can be high in sugar.

3/4 cup coconut milk
1 large egg
1/2 cup coconut nectar or liquid honey
2 tbsp melted coconut oil
1/2 tsp pure vanilla extract
1/2 cup all-purpose flour
1/2 cup coconut flour
1/4 cup Coconut Granola (page 37)
1/4 cup grated fresh or dried coconut
2 tsp baking powder
1/4 tsp salt
1 cup blueberries, fresh or frozen (drained if frozen)

1. Preheat oven to 400°F. Line a 12-cup muffin pan with paper liners or grease the cups with coconut oil.
2. In a bowl, whisk together milk, egg, nectar, oil and vanilla.
3. In a large bowl, combine all-purpose flour, coconut flour, granola, grated coconut, baking powder and salt. Stir well.
4. Make a well in the dry ingredients and stir the milk mixture into it, mixing just until moistened. Fold in the blueberries.
5. Pour the mixture evenly into prepared muffin cups and bake in preheated oven for 15 minutes or until a tester inserted into the centre of a muffin comes out clean.

Coconut Quinoa

Makes 4 to 6 servings

Protein and healthy fats are the key to a perfect breakfast. Nothing can touch quinoa for its high protein content and nutty flavour. Don't forget to rinse the quinoa before cooking to prevent it from being bitter. For a quick breakfast, make this dish ahead of time and warm it up in the morning. (It also makes a great mid-afternoon snack or side dish with fish or chicken.)

- 1 cup quinoa, rinsed and drained
- 1 can (14 oz/398 mL) coconut milk
- 1/2 cup coconut water or plain water
- 1 tbsp softened coconut oil
- 1/4 cup shredded coconut, fresh or dried
- 1/4 cup chopped dried cherries or cranberries (optional)

1. Using a fine-mesh sieve, rinse quinoa under cool running water.
2. In a saucepan, combine quinoa, milk, coconut water and oil. Bring to a boil over high heat. Cover, reduce heat to low and cook for 10 minutes. Turn off heat and leave the covered saucepan on the burner for 6 minutes.
3. Stir in coconut and cherries (if using). Store quinoa in an airtight container in the refrigerator for up to 2 days.

You can omit the coconut oil, but it is the ingredient that will give this breakfast cereal staying power. Add chopped nuts, seeds, berries and tree fruit, as desired.

Roasted Vegetable Frittata

Makes 4 to 6 servings

If you're having company for the weekend, this is a great brunch or breakfast dish that you can make the night before. You can use almost any vegetable you have in the fridge or freezer and it will turn out fabulously. Coconut milk gives the frittata an exotic taste and brings out the flavours of the vegetables.

- 1 onion, cut into quarters
- 1 red bell pepper, cut into 2-inch pieces
- 1 small eggplant, cut into 1-inch cubes
- 1 leek, white and light green parts, cut into 1-inch pieces
- 2 tbsp melted coconut oil
- 6 large eggs
- 1/3 cup coconut milk
- 1/2 cup shredded cheddar cheese
- 3 slices stale whole wheat bread, cut into 1-inch cubes
- 1 tbsp chopped fresh basil
- 2 tsp fresh thyme leaves
- Sea salt and freshly ground black pepper

1. Preheat oven to 375°F. Lightly oil a 9-inch square baking dish.
2. In prepared baking dish, combine onion, red pepper, eggplant and leek. Add oil and toss to coat well. Roast in preheated oven for 25 minutes or until vegetables are tender and slightly charred at the edges. Set baking dish aside to cool. Do not turn off oven.
3. Meanwhile, in a bowl, beat together eggs and milk. Add cheese, bread, basil, thyme and salt and pepper. Pour over roasted vegetables. (Mixture may be prepared up to this point and refrigerated several hours ahead of time or overnight. Return to room temperature before baking.)
4. Bake in preheated oven for 25 to 30 minutes or until lightly browned and set. Set aside to cool for 10 minutes before cutting. Serve warm or at room temperature.

Frozen Breakfast Smoothie

Makes 6 frozen pops

The fibre from the fruit and rolled oats combined with the coconut oil in this smoothie will keep you feeling full all morning, carrying you through to lunchtime or beyond. As a frozen treat, it makes a fast, healthy snack—perfect morning, noon or night as a grab-and-go option kids can easily get for themselves.

- 1 cup chilled pure coconut water
- 1 cup vanilla-flavoured Greek-style yogurt
- 1 cup frozen mango chunks
- 2 tbsp grated orange rind
- 2 oranges, separated into segments, seeds removed
- 1 tbsp large-flake rolled oats
- 1 to 3 tbsp softened coconut oil

1. In a blender, combine coconut water, yogurt, mango, orange rind, orange segments, rolled oats and oil. Secure lid and blend until smooth. Spoon into frozen pop containers and freeze for at least 4 hours or overnight.

Frozen mango chunks make this smoothie or frozen pop very thick, but you can use fresh mango, if desired. Try 1 cup frozen berries in place of the mango. You can substitute instant oats for the large-flake rolled oats.

If you're new to using coconut oil, you may wish to start with 1 teaspoon of coconut oil and gradually increase to 2 tablespoons over time.

Green Energy Smoothie

Makes 1 to 2 servings

Coconut milk gives a touch of sweetness and takes the bitter edge off the greens in this delicious smoothie. I prefer to use frozen greens in smoothies for their convenience. Note that some raw greens such as kale and spinach can interfere with thyroid function, so it's always good to steam or blanch them before you eat them.

1-1/2 cups coconut milk
2 tbsp sunflower seeds
1 to 2 tbsp softened coconut oil
1 tbsp chia seeds or ground flaxseeds
2 cups chopped spinach or kale, fresh or frozen
2 ripe pears, cored and quartered

1. In a blender, combine milk, sunflower seeds, oil, chia seeds, spinach and pears. Secure lid and blend until smooth.

Chia seeds (*Salvia hispanica*) are a great source of healthy omega-3 fats and fibre. When soaked in liquid, they turn thick and gel-like—perfect in smoothies.

If you're new to using coconut oil, you may wish to start with 1 teaspoon of coconut oil and gradually increase to 2 tablespoons over time.

Strawberry Day Starter Smoothie

Makes 1 to 2 servings

Smoothies are a fantastic vehicle for getting your hit of coconut oil. Try to add 1 or 2 tablespoons of coconut oil to all of your smoothies—it acts as a wick to draw out and burn fats that are stored in your body. This is my tropical spin on the basic California smoothie that started the whole smoothie craze: orange juice, banana and strawberries.

1 cup coconut milk
1/4 cup orange juice
1/4 cup thick or strained plain yogurt
1 or 2 tbsp softened coconut oil
1 cup strawberries, fresh or frozen
1 orange, separated into segments
1 banana, cut into 1-inch pieces

1. In a blender, combine milk, orange juice, yogurt, oil, strawberries, orange and banana. Secure lid and blend until smooth.

If you're new to using coconut oil, you may wish to start with 1 teaspoon of coconut oil and gradually increase to 2 tablespoons over time.

3

Lunch

Salads

Soups and Stews

Brocco-Cauliflower Salad with Honey Lime Coconut Dressing

Makes 6 servings

The Honey Lime Coconut Dressing on this salad really complements and sweetens the cruciferous vegetables. People who don't like Brussels sprouts or cauliflower may be surprised. Garnish this colourful salad with Coconut Gremolata (page 94). It's also great topped with crispy bacon bits, toasted coconut or sesame seeds.

2 lb (1 kg) mini red potatoes
1 head broccoli, cut into spears
1/2 head cauliflower, cut into florets
8 to 10 small Brussels sprouts
1/2 cup Honey Lime Coconut Dressing (page 146)
1/2 red onion, thinly sliced

1. In a large saucepan, bring 4 cups of salted water to a boil. Add potatoes and simmer for 10 minutes. Add broccoli, cauliflower and Brussels sprouts. Simmer for 6 minutes or until vegetables are tender-crisp. Drain and rinse with cold water.
2. In a bowl, toss prepared vegetables with dressing. Add onion and stir to combine.

Both broccoli and cauliflower produce florets as they mature. To prepare broccoli, I usually use 2 to 3 inches of the stem and cut the florets into halves or quarters to make a spear shape. Cauliflower is cored and the florets are cut off the centre of the stalk, leaving only a 1/2- to 1-inch stem. If the floret is big, I cut it into halves or quarters.

You can prepare the dressing while the vegetables are cooking. I always make the full cup of dressing (even though this recipe only uses 1/2 cup) because it comes in handy as a substitute for mayonnaise, as a sauce for pasta or as a dressing for green salads.

Cherry Quinoa Salad

Makes 4 servings

A few years ago, I was in Midland, Michigan, and brought home five pounds of dried cherries that were absolutely stunning. Their brilliant colour and sweet citrus flavour can enhance so many dishes and work beautifully with the quinoa, coconut and salty creaminess of the pistachios in this salad.

- 1 cup quinoa, rinsed and drained
- 2 tbsp melted coconut oil
- 1 onion, finely chopped
- 3 tbsp coconut vinegar or balsamic vinegar
- 1/4 cup dried cherries
- 1/4 cup coarsely chopped pistachio nuts
- 1/4 cup chopped fresh parsley
- Sea salt and freshly ground black pepper

1. In a saucepan, bring 2 cups of water to a boil. Add quinoa, cover and reduce heat to low. Simmer for 15 minutes or until water has been completely absorbed and quinoa is tender. Fluff with a fork, let stand for 5 minutes and transfer to a salad bowl. Set aside.
2. Meanwhile, in a skillet, heat oil over medium heat. Sauté onion for 5 minutes or until soft. Remove from heat and stir in vinegar.
3. Add onion mixture, cherries, pistachios and parsley to quinoa and toss to combine. Season to taste with salt and pepper.

Citrus Cabbage Salad with Orange Coconut Dressing

Makes 6 servings

Here's a new twist on an old favourite, the Waldorf salad. The Orange Coconut Dressing really makes this version come alive.

- 4 cups thinly sliced cabbage
- 2 cups shredded carrot
- 2 apples, finely chopped
- 1/2 cup coarsely chopped dates
- 1/2 cup coarsely chopped walnuts
- 1/4 cup shredded unsweetened coconut, fresh or dried
- 1 cup Orange Coconut Dressing (page 152)

1. In a salad bowl, combine cabbage, carrot, apples, dates, walnuts and coconut. Drizzle with dressing and toss to coat well.

You'll find several interesting varieties of cabbage at farm stands and farmers' markets. I like to use red cabbage in this recipe, but you can experiment with Savoy, pictured left, or bok choy.

Curried Coconut Lettuce Wraps

Makes 8 wraps

These small lettuce packets make a neat appetizer or salad alternative at lunch or dinner. You can make them up to a day ahead and keep them, tightly covered, in the refrigerator.

- 2 tbsp melted coconut oil
- 1 onion, chopped
- 2 cloves garlic, finely chopped
- 1 small jalapeno pepper, finely chopped (optional)
- 1 tbsp Madras Curry Spice Blend (page 139)
- 1/2 cup Coconut Garlic Mayonnaise (page 155), Coconut Yogurt Mayonnaise (page 156) or regular mayonnaise
- 1/4 cup plain yogurt
- 1 sweet potato, cooked and diced
- 1/3 cup chopped almonds
- 1/4 cup toasted shredded coconut, fresh or dried
- 8 lettuce leaves

1. In a skillet, heat oil over medium heat. Sauté onion for 5 minutes. Add garlic, jalapeno (if using) and curry spice blend. Cook, stirring constantly, for 2 minutes. Remove from heat and set aside to cool.
2. Meanwhile, in a large bowl, combine mayonnaise, yogurt, sweet potato, almonds and toasted coconut. Add cooled onion mixture and stir well. Scoop 1/3 cup of the filling mixture into the centre of a lettuce leaf. Fold the top and bottom of the lettuce over the filling and roll into a wrap. Repeat with remaining lettuce and filling.

You can substitute a pinch to 1-1/4 tsp cayenne pepper or any fresh hot pepper for the jalapeno.

For this recipe, use long-strand or thick flakes of sweetened or unsweetened coconut. (Note that sweetened coconut will toast faster than unsweetened coconut.)

To toast coconut: Preheat oven to 350°F. Arrange coconut in an even layer on a rimmed baking sheet and toast, stirring once or twice, in preheated oven for 5 to 7 minutes or until lightly brown.

Use iceberg or Boston bibb lettuce for the wraps. Wash and core the lettuce. Separate, dry and chill the leaves in the refrigerator for 30 minutes or until crisp.

Mango Chicken Salad

Makes 4 servings

Don't be turned off by the number of ingredients in this recipe. It's really not that hard to make, especially if you cook the chicken breast ahead of time.

- 1 boneless skinless chicken breast
- 2 cups chicken broth or water
- 1 tbsp coconut vinegar
- 1 tbsp coconut nectar
- 1 tbsp melted coconut oil
- 1 onion, finely chopped
- 1 clove garlic, minced
- 2 tbsp Madras Curry Spice Blend (page 139)
- 2 tbsp coconut flour
- 1/2 cup coconut milk
- 1/2 cup Coconut Garlic Mayonnaise (page 155), Coconut Yogurt Mayonnaise (page 156) or regular mayonnaise
- 4 mangoes, diced
- 2 stalks celery, chopped
- 1 cup fresh bean sprouts
- Sea salt and freshly ground black pepper
- 4 chilled lettuce leaves

1. In a saucepan, combine chicken, broth, vinegar and nectar. Bring to a boil over high heat. Cover, reduce heat and gently simmer for 8 to 12 minutes or until chicken is cooked through and reaches 165°F on a meat thermometer. Transfer chicken to a plate and set aside to cool. Reserve broth in the pan.
2. Meanwhile, in a skillet, heat oil over medium heat. Sauté onion for 5 minutes. Add garlic and curry spice blend and cook, stirring frequently, for 1 minute. Add flour and cook, stirring constantly, for 1 minute. Whisk in reserved chicken poaching liquid and cook, whisking constantly, for 6 minutes or until mixture begins to thicken. Add milk and cook, stirring frequently, for 3 minutes or until mixture is thick and creamy. Set aside to cool. Stir in mayonnaise.
3. Meanwhile, dice or shred poached chicken breast and transfer to a large bowl. Add mangoes, celery, bean sprouts and mayonnaise mixture and toss to combine. Season to taste with salt and pepper. Wash, core and drain lettuce well. Pat dry, cover and chill in the refrigerator for at least 30 minutes or up to a day in advance. To serve, divide salad into 4 equal portions and spoon into lettuce cups.

Large leaf lettuce such as Boston bibb or iceberg work best in this recipe if you scoop the salad into cups. You can also serve on a bed of mixed greens instead.

For convenience, you can make the salad mixture up to 2 days in advance and store, tightly covered, in the refrigerator. Fill lettuce cups just before serving.

Lettuce Wedges with Coconut Blue Cheese Dressing

Makes 4 servings

This salad is a slice of pure, cool, crisp pleasure—perfect for hot summer days. It's no wonder a version of this salad was a standard in the 1950s. After using artisanal greens for so long, I was pleasantly surprised by the sweetness of iceberg lettuce, which has been languishing in produce aisles for a long time. The real secret is to chill the lettuce well.

- 1 head iceberg lettuce
- 1 cup Coconut Blue Cheese Dressing (page 146)
- 1/2 cup chopped pecans or walnuts
- 1/2 cup crisp bacon bits (optional)
- 1/2 cup chopped dried cherries or cranberries

1. Wash, core and drain lettuce well. Pat dry, cover and chill in the refrigerator for at least 30 minutes or up to 1 day in advance.
2. Cut the lettuce into 4 equal wedges. Drizzle each wedge with about 1/4 cup dressing and sprinkle with equal amounts of pecans, bacon (if using) and cherries.

If using a large head of lettuce, you may wish to halve the head and then cut 3 wedges out of each half (instead of 2) so the wedges are smaller.

Roasted Beet and Pineapple Salad with Honey Lime Coconut Dressing

Makes 4 servings

Pineapple is a surprising ingredient here, but it takes the dish out of the realm of the ordinary. It's my default salad in the winter.

- 2 lb (1 kg) baby beets, trimmed
- 2 tbsp melted coconut oil
- 2 cups drained pineapple chunks, fresh or canned
- 1/4 cup shredded fresh basil
- 1/2 cup crumbled feta cheese or soft goat cheese
- 1/2 cup Honey Lime Coconut Dressing (page 146)

1. Preheat oven to 400°F. Line rimmed baking sheet with parchment paper.
2. On prepared baking sheet, toss beets with oil. Roast in preheated oven for 30 minutes. Stir in pineapple and roast for another 15 minutes or until beets are easily pierced with a knife. Set aside to cool. Peel beets if required and cut into quarters.
3. In a bowl, combine cooked beets and pineapple, basil and feta. Drizzle with dressing and toss to coat well.

Baby beets are small and tender and the skin has not yet thickened, so they usually do not require much in the way of trimming. To prepare them for this dish, using a sharp knife, cut off the top and the stringy tail and scrape away any thick skin around the top of the beet; rinse and pat dry.

If using mature beets with thick, coarse skins, trim as described above, cut into quarters, roast, cool and either rub off the skin with your fingers or use a paring knife to peel it off.

Don't throw away the tasty beet greens: Wash, dry and use them in salads and recipes wherever kale or spinach is called for.

Southwestern Cauliflower and Potato Salad

Makes 4 servings

Try making your own spice blends—they are the cornerstone of most recipes. This dish is the perfect example: Coconut Jerk Spice Rub delivers the essence of Southwestern flavour. You might think smokey chile with coconut is an unusual pairing, but it works.

- 2 lb (1 kg) mixed mini red and yellow potatoes
- 1/2 small head cauliflower, cut into florets
- 4 tbsp melted coconut oil, divided
- 1 onion, quartered
- 2 cloves garlic, finely chopped
- 1 red or orange bell pepper, coarsely chopped
- 1 tbsp Coconut Jerk Spice Rub (page 134)
- 1 tbsp whole fennel seeds
- 1 tsp whole cumin seeds
- 1/4 cup toasted sunflower seeds

1. In a saucepan, bring 6 cups water to a boil over high heat. Add potatoes and return to a boil. Reduce heat and simmer for 15 minutes. Add cauliflower and simmer for 5 to 7 minutes or until tender. Drain and rinse under cool running water. Set aside.
2. In a skillet, heat 2 tbsp oil over medium heat. Sauté onion for 5 minutes. Add garlic, orange pepper, spice rub, fennel and cumin and cook, stirring frequently, for 3 minutes or until peppers are tender. Add remaining oil and potatoes, cauliflower and sunflower seeds and cook, stirring frequently, for 3 to 5 minutes or until potatoes and cauliflower are lightly browned. Serve warm or chilled.

Asian Greens Soup

Makes 4 to 6 servings

On its own, with the greens and coconut milk, this is a warm and comforting soup, but sometimes I add cooked noodles, rice or other grains if I have leftovers. Vietnamese cinnamon is different than the variety we normally get in North America—its warming sweetness, along with the coriander, is what really gives this soup its Asian flavour.

- 2 tbsp melted coconut oil
- 1 onion, chopped
- 3 cloves garlic, finely chopped
- 1 tbsp Vietnamese Five-Spice Blend (page 143)
- 3 cups chicken broth
- 2 cups thinly sliced bok choy
- 2 cups chopped kale
- 1 cup fresh snow peas, sliced lengthwise
- 1 can (14 oz/398 mL) coconut milk
- 1/4 cup chopped fresh parsley

1. In a soup pot or large saucepan, heat coconut oil over medium heat. Sauté onion for 5 minutes or until soft. Add garlic and spice blend and cook, stirring constantly, for 2 minutes or until garlic is soft.
2. Add broth and bring to a boil. Stir in bok choy, kale and snow peas and simmer for 10 minutes. Stir in coconut milk and parsley and simmer for 2 to 3 minutes or until peas are tender-crisp.

Bok choy and baby bok choy (*Brassica rapa* sb. *Pekinensis*) or other varieties of Chinese cabbage are all suitable for use in this healthy soup. Dice, chop or thinly slice the tender white stems and chop or thinly shred the leaves. Remove and discard any tough stems and ribs from the leaves.

Snow peas (*Pisum sativum* var. *saccharatum*), often called sugar snap peas, are the fresh, immature and tender peas in the pod most often used in Asian dishes. Available most of the year, the whole pod may be chopped or thinly sliced lengthwise and enjoyed raw or cooked in this recipe.

Kale (*Brassica oleracea* acephala group) is enjoying a well-deserved popularity due to the vitamins A, B and C, calcium and magnesium it contains. If you are just starting to experiment with kale, look for young leaves—they are more tender and slightly sweeter. If using mature kale, remove and discard tough stems and ribs, stack the leaves, and use a sharp knife to chop or shred them. You can find heirloom varieties of kale at many farmers' markets. Purple kale, Ragged Jack and Black Tuscan are my favourites.

Fish Chowder

Makes 4 to 6 servings

In the summer, when tomatoes are fresh and fabulous, I coarsely chop a couple and toss them into this soup for added colour, taste and goodness. If summer squash is not available, substitute equal amounts of zucchini or leek.

3 slices side bacon, chopped
2 tbsp melted coconut oil
1 onion, chopped
2 cloves garlic, finely chopped
1 small yellow summer squash or green zucchini, diced
2 cans (14 oz/398 mL each) coconut milk
2-1/2 cups diced potatoes (about 2 potatoes)
2 stalks celery, diced
2 bay leaves
1 lb (500 g) whitefish, cut into 1-inch cubes
1 can (5.4 oz/160 mL) coconut cream
1/4 cup chopped fresh parsley

1. In a soup pot or large saucepan, cook bacon over medium-low heat, stirring frequently, for 10 minutes or until crisp and all of the fat has been released. Using a slotted spoon, transfer bacon to a plate lined with paper towel and set aside. Drain and discard bacon fat from pot.
2. Using the same pot, heat oil over medium heat. Sauté onion for 5 minutes or until soft. Add garlic and squash and cook, stirring frequently, for 5 minutes. Add milk and bring to a boil. Stir in potatoes, celery and bay leaves. Cover, reduce heat to low and simmer for 20 minutes. Add whitefish, reserved bacon, cream and parsley and simmer for 10 to 15 minutes or until potatoes are fork-tender and fish is cooked through (when done, the fish will turn opaque and flake easily with a fork).

Small, tender yellow summer squash or zucchini work best in this recipe.

In place of whitefish, try making this chowder with fresh or frozen snapper, cod, halibut, trout or salmon. You can also add scallops, shrimp or crab.

Gingered Carrot Squash Soup

Makes 6 servings

This soup is a great winter standby that's filling and warming (ginger is a warming herb).

- 2 tbsp melted coconut oil
- 1 onion, chopped
- 1 tbsp Garam Masala Spice Blend (page 137)
- 1 tbsp finely chopped gingerroot
- 3 cups chicken or vegetable broth
- 1 carrot, diced
- 1/2 butternut or acorn squash, peeled, seeded and diced
- 1/2 cup coconut cream, divided
- Sea salt and freshly ground black pepper
- 3 tbsp unsweetened coconut flakes, fresh or dried

1. In a soup pot or large saucepan, heat oil over medium heat. Sauté onion for 5 minutes or until soft. Add Garam Masala Spice Blend and ginger and cook, stirring constantly, for 1 minute. Add broth and bring to a boil. Stir in carrot and squash. Reduce heat and simmer for 20 minutes or until vegetables are tender.
2. Add cream and, using an immersion blender, purée the soup. Season to taste with salt and pepper. To serve, garnish with toasted coconut flakes.

If you don't have an immersion blender, use a regular blender: In 2 batches, purée the soup, adding 1/4 cup coconut cream to each batch. Return puréed soup to pot over medium heat and stir for 1 minute or until heated through.

Three Bean Chili

Makes 6 to 8 servings

Every good cookbook needs a chili recipe—this is mine. I like this dish because it's proof that you don't need meat to have a hearty, warming meal. Garnish the chili with shredded cheddar cheese, Coconut Gremolata (page 149) or chopped, toasted nuts. Serve with warm tortillas or crusty country bread.

- 2 tbsp coconut oil
- 1 onion, chopped
- 1 red or green bell pepper, chopped
- 2 cloves garlic, finely chopped
- 1 tbsp Coconut Jerk Spice Rub (page 134)
- 1 can (28 oz/796 mL) diced tomatoes and juices
- 1 can (14 oz/398 mL) coconut milk
- 1 tbsp coconut aminos or tamari sauce
- 1 can (14 oz/398 mL or 19 oz/540 mL) pinto beans, rinsed and drained
- 1 can (14 oz/398 mL or 19 oz/540 mL) black beans, rinsed and drained
- 1 can (14 oz/398 mL or 19 oz/540 mL) red kidney beans, rinsed and drained
- 1 tbsp chopped fresh oregano
- 1/4 cup chopped fresh cilantro or flat-leaf parsley

1. In a large saucepan or Dutch oven, heat oil over medium heat. Sauté onion and bell pepper for 5 minutes. Add garlic and spice rub and cook, stirring frequently, for 2 minutes. Add tomatoes and juices and bring to a boil.
2. Add milk, aminos, pinto beans, black beans, kidney beans and oregano. Cover, reduce heat to low and simmer, stirring occasionally, for 20 minutes. Stir in cilantro.

If desired, you can substitute Coconut Jerk Spice Rub (page 134) or store-bought Jamaican jerk spice for the Coconut Jerk Spice Rub.

You can use chickpeas, lentils, lima beans or white beans in this recipe. If you prefer a meaty chili, sauté 1/2 lb (250 g) ground chicken or pork with the onion in Step 1.

A Dutch oven is a large cooking pot with a lid used to simmer foods on top of the stove or in the oven for long periods of time. It is traditionally made of cast iron or ceramic-coated cast iron, but stainless steel versions are available. They are heavy, and the lids are usually dimpled in order to condense steam and return the liquids to the pot. If you don't have a Dutch oven, you can use a large cast iron skillet or a pot, with a lid, that is both flame- and oven-proof.

Creamy Asparagus Soup

Makes 4 servings

I live close to an asparagus farm, so when asparagus is in season, we practically live on it. If you're like me and you devour asparagus when it's fresh in the spring, this soup is a great way to use up the stalks, and you can save the tips for more flashy stir-fry dishes. Coconut milk is a convenient and healthier substitute for heavy cream, and its subtle sweetness works well in savoury dishes.

- 1 tbsp melted coconut oil
- 1 onion, chopped
- 1 leek, white and light green parts, chopped
- 1 tbsp fresh thyme leaves
- 1 potato, chopped
- 3 cups chicken broth
- 2 lb (1 kg) fresh asparagus, trimmed and cut into 1-inch pieces
- 1 can (5.4 oz/160 mL) coconut cream
- 1/2 cup toasted coconut or Coconut Gremolata (page 149)

1. In a soup pot or large saucepan, heat oil over medium heat. Sauté onion and leek for 5 minutes or until soft. Add thyme and potato and cook, stirring constantly, for 1 minute. Add broth and bring to a boil. Reduce heat and simmer for 15 minutes. Stir in asparagus and cream and simmer for 5 minutes or until vegetables are tender.
2. To serve, garnish with toasted coconut.

Most people assume that young, thin stalks of asparagus will be more tender than older, thicker stalks, but in fact, thicker stalks are best. The cellulose (the substance that makes it possible for the plant to shoot out of the ground and stand straight up) is stronger—and tougher—in thin stems than in larger stalks that have fleshed out.

Spiced Lamb and Peanut Stew

Makes 4 to 6 servings

I love this combination of spiced lamb with coconut cream and peanuts. It's a delicious dish for a humble stew—you can use this one for entertaining anytime.

- 2 tbsp melted coconut oil
- 2 onions, chopped
- 2 cloves garlic, finely chopped
- 1 fresh or dried cayenne pepper, finely chopped (optional)
- 1 tbsp Madras Curry Spice Blend (page 139)
- 1-1/2 lb (750 g) boneless lamb, cut into 1-inch cubes
- 1 can (28 oz/796 mL) diced tomatoes and juices
- 4 carrots, sliced
- 2 potatoes, cut into 1-inch cubes
- 1 can (5.4 oz/160 mL) coconut cream
- 1/4 cup coarsely chopped roasted peanuts
- 1/4 cup chunky peanut butter

1. In a soup pot or large saucepan, heat oil over medium heat. Sauté onions for 5 minutes. Add garlic, cayenne pepper (if using) and curry spice blend and cook, stirring constantly, until garlic is soft, 2 to 3 minutes. Add lamb and cook, stirring constantly, until evenly browned, 2 to 3 minutes. Stir in tomatoes with juices and bring to a boil. Reduce heat to low, cover and simmer for 50 minutes or until lamb is cooked and tender. Add carrots and potatoes and simmer for 10 minutes or until fork-tender.
2. Add cream, peanuts and peanut butter and stir to combine. Continue to heat until last 3 ingredients are warm.

Madras Curry Spice Blend is mildly hot to my palate. If desired, omit the cayenne pepper and use only 1 tsp of the curry spice blend. You can always add more heat and spice to suit your taste.

Vegetable Hot Pot

Makes 4 servings

This makes for a quick meal. If your kids don't like the veggies in the recipe, you can swap them with others they prefer. For a fun family lunch, kids can cook their own vegetables on skewers if you have a portable hotplate to keep the broth simmering.

- 2 tbsp coconut oil
- 1 onion, cut into quarters
- 2 long, thin Japanese eggplants, cut into 1-inch cubes
- 2 cloves garlic, finely chopped
- 1 tbsp Madras Curry Spice Blend (page 139)
- 2 cups chicken broth
- 2 parsnips, cut into 1-inch cubes
- 2 tbsp creamed coconut
- 1 tbsp freshly squeezed lime juice
- 1/4 cup shredded fresh basil or Coconut Gremolata (page 149)

1. In a large saucepan or Dutch oven, heat oil over medium heat. Sauté onion and eggplant for 5 minutes. Add garlic and curry spice blend and cook, stirring frequently, for 2 minutes. Add chicken broth and bring to a boil.
2. Add parsnips, cover and reduce heat to low. Simmer, stirring occasionally, for 10 minutes or until vegetables are tender. Stir in creamed coconut, lime juice and basil and heat through.

A Dutch oven is a large cooking pot with a lid used to simmer foods on top of the stove or in the oven for long periods of time. It is traditionally made of cast iron or ceramic-coated cast iron, but stainless steel versions are available. They are heavy, and the lids are usually dimpled in order to condense steam and return the liquids to the pot. If you don't have a Dutch oven, you can use a large cast iron skillet or a pot, with a lid, that is both flame- and oven-proof.

4

Dinner

Main Dishes

Noodles and Pasta

Vegetable Side Dishes

Braised Beef and Vegetables in Coconut Sauce

Makes 4 to 6 servings

Because you are braising, you can use a less-tender cut, but be sure to pound the meat to help tenderize it. The vegetables cook into the coconut sauce and give it a rich flavour and texture.

- 1 lb (500 g) flank steak or inside beef round
- 2 tbsp coconut flour, divided
- 1 tbsp coconut oil
- 1 onion, coarsely chopped
- 3 cloves garlic, finely chopped
- 1 eggplant, coarsely chopped
- 1 red bell pepper, coarsely chopped
- 3/4 cup coconut milk
- 1 tbsp coconut aminos or tamari sauce

1. Place steak on a large sheet of parchment paper or cutting board. Sprinkle with 1 tablespoon of flour and, using a meat mallet, pound until about 1/2 inch thick (or thinner than it was when you started). Flip and repeat with remaining flour. Cut into 1/2-inch-wide strips. Cut the strips into 1- to 2-inch lengths. Set aside.
2. In a flameproof tagine, Dutch oven or large skillet, heat oil over medium heat. Add onion and sauté for 5 minutes. Add garlic and cook, stirring frequently, for 1 minute. Add beef strips and cook, stirring constantly, for 3 to 4 minutes or until browned on all sides. Stir in eggplant, red pepper, milk and aminos. Bring to a boil. Cover, reduce heat to low and simmer for 20 minutes or until meat is tender and vegetables have been reduced into the sauce.

You can use 2 or 3 small, thin Japanese eggplants (photograph, at left) or 1 plump Italian eggplant (photograph, at right).

To prepare the eggplant: Using a sharp knife, remove the top and bottom from the eggplant. Partially remove the skin in alternating 1-inch strips to make a striped pattern. Stand the eggplant up on its wide end and cut it lengthwise into 1/4-inch slices. Arrange on the cutting board in stacks of 3. Cut each stack lengthwise into 1/2-inch strips, then cut across the strips to make 1/2-inch cubes.

I use my tagine for this dish, but you can use a heavy-bottomed skillet or Dutch oven as long as it has a lid. Be sure that when you turn the heat down to simmer in Step 2, the mixture actually does simmer. You may have to check after a few minutes and adjust the heat up or down in order to keep the mixture simmering just under a hard boil

Bean and Nut Loaf

Makes 6 servings

This dish made its debut at an annual meeting of the Women's Culinary Network in Toronto and was well-received by many discerning Canadian foodies. It raises the question "Who needs meat?" Obviously no one when you combine mushrooms, cashews and coconut milk with a great spice blend. The loaf holds together beautifully and is delicious served warm or at room temperature the next day. I'm sure it will be just as much of a hit at your next potluck outing.

- 1 tbsp melted coconut oil
- 1 onion, chopped
- 2 cups chopped mushrooms
- 1 tbsp Garam Masala Spice Blend (page 137)
- 1 to 2 dried chiles, crushed (optional)
- 1/4 cup coconut milk
- 3 large eggs
- 3/4 cup fresh whole wheat breadcrumbs
- 1/4 cup coconut flour
- 1 can (19 oz/540 mL) white beans, rinsed, drained and mashed
- 1-1/2 cups cashews, chopped
- 1/3 cup chili sauce

1. Preheat oven to 350°F. Line an 8- × 4-inch loaf pan with parchment paper or grease lightly with coconut oil.
2. In a skillet, heat oil over medium heat. Sauté onion for 5 minutes. Stir in mushrooms, spice blend and chile (if using). Cook, stirring frequently, for 6 minutes or until vegetables are tender. Add milk and simmer, stirring occasionally, for 10 minutes or until liquid is reduced. Set aside to cool.
3. In a large bowl, beat together eggs, breadcrumbs and flour. Add mushroom mixture, beans and cashews and stir until mixed well. Spoon into prepared pan, packing the mixture down with the back of a spoon. Make a well down the centre of the top of the mixture and pour chili sauce into the indentation. Bake in preheated oven for 1 hour or until the loaf begins to pull away from the sides of the pan. Set aside for 15 minutes before lifting out of pan and slicing.

Chicken in Pineapple Boats

Makes 4 servings

If you want drama, this recipe is a showboat, and yet it is so easy to prepare. Of course, you can use canned pineapple chunks and serve it on a dinner dish, but what fun would that be?

- 2 cups strained plain yogurt
- 1/4 cup shredded coconut, fresh or dried
- 1 tbsp freshly squeezed lime juice
- 1 tbsp Tandoori Spice Blend (page 142)
- 4 boneless skinless chicken breasts, cut into 1-1/2-inch cubes (about 1-1/2 lb/750 g)
- 2 ripe pineapples
- 1/4 cup toasted slivered almonds
- 1/4 cup toasted grated coconut, fresh or dried

1. Preheat barbeque to medium-high. Lightly oil 8 metal kebab skewers .
2. In a large bowl, combine yogurt, shredded coconut, lime juice and spice blend. Add chicken and stir to coat well. Cover and refrigerate for 30 minutes.
3. Meanwhile, using a sharp knife, cut pineapples in half through the top leaves lengthwise to the base. Using a paring knife, cut out the flesh, leaving the shells intact. Set shells aside. Cut pineapple flesh into 1-1/2-inch cubes. Thread even portions onto 4 prepared skewers. Set aside.
4. Thread even portions of the chicken onto remaining skewers. Grill for 5 minutes. Turn chicken over. Add pineapple skewers to the grill. Grill both chicken and pineapple for 4 to 5 minutes, turning pineapple once, until chicken is cooked through and reaches 165°F on a meat thermometer.
5. Slide chicken and pineapple off skewers into pineapple shells and garnish with almonds and coconut.

You can use wooden skewers, but metal conducts heat best so the food cooks faster.

Grill the pineapple and chicken on separate skewers—the chicken takes longer to cook.

An outdoor barbeque works best, but you can also use a cast iron grill pan or the oven broiler (cooking times may vary).

To strain yogurt: Line a sieve with a double layer of cheesecloth and place over a bowl. Empty one 1-lb (500 g) container of yogurt into the sieve and refrigerate for 15 minutes. (You can let it sit overnight, but the strained yogurt will measure less than the 2 cups called for in the recipe. The longer the yogurt drains, the thicker it becomes.) Discard the liquid or use it to substitute for milk in baked goods or soup.

To toast coconut: Preheat oven to 350°F. Arrange coconut in an even layer on a rimmed baking sheet and toast, stirring once or twice, in preheated oven for 5 to 7 minutes or until lightly brown.

Citrus Glazed Turkey

Makes 4 servings

The sweet but tart glaze is exceptional on poultry, fish and pork. Brush it on liberally while grilling the meat, but be sure to discard any unused portions.

Glaze

Makes 2-1/4 cups

1 cup coconut nectar or liquid honey

1/2 cup freshly squeezed lemon juice

1/2 cup freshly squeezed orange juice

1/4 cup tamari or soy sauce

Rub

Makes 1/4 cup

3 tbsp Coconut Jerk Spice Rub (page 134)

1 tbsp coconut flour

4 skinless turkey thighs (about 2 lb/1 kg)

1. In a saucepan, combine nectar, lemon juice, orange juice and tamari. Bring to a boil over high heat, reduce heat and simmer for 5 minutes or until thickened slightly. Transfer to a large bowl and set aside to cool.
2. In a small bowl, combine spice rub and coconut flour. Using your hands, rub the mixture evenly onto the turkey thighs, coating both sides. Transfer the thighs and any loose rub to the cooled glaze in the saucepan. Cover and refrigerate for at least 30 minutes or overnight. Bring to room temperature before cooking.
3. Preheat barbeque to medium-high.
4. Using tongs, transfer the thighs to the preheated grill, reserving glaze. Cook, turning often and brushing with reserved glaze, for 15 minutes or until the turkey is cooked through and reaches 160°F on a meat thermometer.

If desired, you can substitute the Coconut Jerk Spice Rub with an equal amount of Jamaican jerk spice, which you can find in the spice aisle of most supermarkets.

Coconut Curry Chicken

Makes 4 servings

Coconut and curry are a dynamite flavour combination, and I use them together a lot. In this dish, the coconut flour coating thickens the coconut milk in the sauce and the curry and ginger spices give it a world-class flavour.

- 2 tbsp coconut flour
- 1 tsp coconut sugar crystals
- 1 tsp sea salt
- 8 chicken thighs (about 2 lb/1 kg)
- 2 tbsp melted coconut oil
- 2 onions, coarsely chopped
- 2 cloves garlic, finely chopped
- 1 tbsp freshly grated gingerroot
- 3 tbsp Madras Curry Spice Blend (page 139)
- 1 cup chicken broth
- 1 cup coconut milk
- 1 banana, coarsely chopped
- 1 sweet potato, coarsely chopped
- 1 bay leaf

1. In a flat bowl or pie plate, combine flour, sugar crystals and salt. Dredge thighs in the mixture to coat them. Reserve any extra flour mixture.
2. In a Dutch oven, heat oil over medium heat. Add chicken thighs and cook for 10 minutes or until browned well on all sides. Transfer to a plate and set aside.
3. Drain all but 2 tablespoons of oil in the pot. Add onions and sauté for 5 minutes. Add garlic, ginger and curry spice blend and cook, stirring constantly, for 2 minutes or until onions are soft. Return reserved chicken to the pot along with any reserved flour mixture. Add broth and milk. Bring to a boil, reduce heat to low and add banana, sweet potato and bay leaf. Cover and simmer for 25 minutes or until potato is fork-tender and chicken is cooked through and reaches 160°F on a meat thermometer.

A Dutch oven is a large cooking pot with a lid used to simmer foods on top of the stove or in the oven for long periods of time. It is traditionally made of cast iron or ceramic-coated cast iron, but stainless steel versions are available. They are heavy, and the lids are usually dimpled in order to condense steam and return the liquids to the pot. If you don't have a Dutch oven, you can use a large cast iron skillet or a pot, with a lid, that is both flame- and oven-proof.

Crispy Baked Fish

Makes 4 servings

The crispy topping on this fish, with its fresh flavours of lemon and coconut, will appeal to kids and those who don't like the taste of fish. It's also great on baked pasta dishes, roasted vegetables and casseroles.

One 1-lb (500 g) fillet of trout, white-fish or salmon
1/2 lemon

Crispy Topping

1/3 cup whole almonds or pecans
3 cloves garlic
1 cube candied ginger
2 slices stale whole wheat bread, cut into 1-inch cubes
2 tbsp softened coconut oil
1 tbsp fresh thyme leaves or chopped fresh basil

1. Preheat oven to 350°F. Line a rimmed baking sheet with parchment paper.
2. Rinse fish under cool running water, pat dry and arrange, skin-side down, on prepared baking sheet.
3. In a food processor or blender, combine almonds, garlic, ginger, bread, oil and thyme and pulse until finely chopped. Spread evenly over the top of the fish, patting to form a crust. Bake in preheated oven for 15 minutes or until the topping is crispy and the fish turns opaque and flakes easily with a fork.

Moroccan Mango Chicken Tagine

Makes 4 servings

Coconut and garam masala make a perfect combination for this fruity chicken dish. Cooking in a tagine is like poaching. The liquids do not evaporate and the fruit reduces into the sauce as the dish cooks.

- 3 tbsp melted coconut oil
- 2 onions, cut into quarters
- 4 cloves garlic, coarsely chopped
- 1 tbsp Garam Masala Spice Blend (page 137)
- 2 dried cayenne peppers, crushed
- 1 tomato, cut into quarters
- 2 mangoes, coarsely chopped
- 1/4 cup coconut milk
- 4 skinless chicken breasts (bone in or out)

1. In a large flameproof tagine or Dutch oven, heat oil over medium heat. Add onions and cook, stirring frequently, for 5 minutes. Add garlic and cook, stirring frequently, for 1 to 2 minutes or until onion is soft. Stir in Garam Masala Spice Blend and cayenne.
2. Add tomato, mangoes and milk. Bring to a boil, stirring constantly. Add chicken, meaty side down, sliding it around and moving onions away so the chicken is in direct contact with the bottom of the tagine. Cook for about 3 minutes or until bottom of chicken is browned.
3. Using tongs, turn chicken over. Cover with lid, reduce heat to low and simmer, stirring once, for 35 minutes or until cooked through and thickest part of chicken reaches 165°F on a meat thermometer.

The cayenne pepper (*Capsicum annum* 'cayenne') is a small, hot red pepper that grows to about 1-1/2 to 2-inches long. It is sometimes available fresh in supermarkets or farmers' markets, but more often it is dried and sold whole, flaked (also known as "hot pepper flakes") or powdered. You can substitute the dried cayenne in this recipe with 1 tsp (or less) hot pepper flakes or cayenne powder.

A tagine is a 2-piece clay pot with a cone-shaped lid that has been used by nomadic Middle Eastern (Berber) tribes for centuries. It is also the word used to describe dishes cooked in the unique pot. Foods cooked in a tagine are incredibly moist, often cooked in velvety fruit sauces that are spectacular when served with rice or noodles. If you don't have a flameproof tagine (one that may be used safely on top of the stove), use a Dutch oven with a lid.

Lamb Curry

Makes 4 servings

Another delicious curry—and a cancer-fighting dish. We rarely think of the foods we eat as helping to reduce the risk of cancer or other diseases, but healthy food is indeed our medicine. This dish has health benefits in spades: coconut oil, onion, garlic and curry spices are all great cancer busters. Serve with couscous, Coconut Rice (page 93) or Herbed Rice Pilaf (page 100).

- 4 tbsp melted coconut oil
- 2 lb (1 kg) boneless lamb, cut into 1-inch cubes
- 1 onion, chopped
- 1 red bell pepper, chopped
- 1 carrot, chopped
- 4 cloves garlic, chopped
- 1 tbsp Madras Curry Spice Blend (page 139)
- 1 tbsp ground coriander
- 1 tsp sea salt
- grated rind of 1 lemon (optional)
- 2 cups coconut milk

1. In a large flameproof tagine or Dutch oven, heat oil over medium heat. Add lamb and cook, turning often, for 4 minutes or until browned on all sides. Transfer to a plate and set aside.
2. Add onion, red pepper and carrot to the tagine and sauté for 5 minutes. Add garlic, curry spice blend, coriander, salt and lemon rind (if using) and cook, stirring frequently, for 1 minute. Add milk and bring to a boil. Add reserved lamb. Cover, reduce heat to low and simmer for 1 hour or until meat is tender.

Mushroom Ragout with Quinoa

Makes 4 servings

This ragout is so versatile: You can also serve it with polenta or scrambled eggs for breakfast or on top of whole wheat toast for lunch. At dinner it makes a substantial meal when teamed with the quinoa and a steamed green vegetable or Bulgur and Lentils with Coconut Gremolata (page 94) or Coconut Barley Risotto (page 96).

- 1 cup dried porcini mushrooms (about 1 oz)
- 2 tbsp melted coconut oil
- 1 lb (500 g) mixed fresh mushrooms, sliced
- 1 onion, finely chopped
- 2 cloves garlic, minced
- 1 tbsp coconut flour
- 1 can (14 oz/398 mL) coconut milk
- 1/4 cup dry white wine or vegetable broth
- 2 tbsp fresh thyme leaves
- 2 cups Coconut Quinoa (page 39)
- 4 sprigs fresh thyme, for garnish (optional)

1. In a bowl, cover the porcini mushrooms with boiling water and set aside to soak for 20 minutes or until mushrooms are rehydrated. Drain (discard or reserve soaking liquid for another recipe) and coarsely chop mushrooms. Set aside.
2. In a skillet, heat oil over medium heat. Add fresh mushrooms, rehydrated mushrooms, onion and garlic and sauté for 8 minutes or until most of the liquid is absorbed. Add flour and stir to coat vegetables. Add milk and cook, stirring constantly, for 8 to 10 minutes or until liquid is thickened. Stir in wine and thyme leaves and cook until heated through.
3. To serve, spoon over Coconut Quinoa and garnish with fresh thyme leaves (if using).

Porcini mushrooms (*Boletus edulis*) are hearty, earthy-tasting mushrooms that are grown in North America and Europe and harvested in autumn. Available both dried and fresh (fresh can be expensive), they are wonderful in soups, stews or grain dishes such as risotto.

You can use one variety of fresh mushrooms or buy a variety that includes cremini, oyster, shiitake, portobello or cep mushrooms.

I like to have a wide variety of dried mushrooms on hand, especially porcini because they are not often available fresh where I live. Save the soaking water and strain it through a coffee filter or double layer of cheesecloth. Store covered in the refrigerator for up to 3 days or in the freezer for up to 3 months and use it in soup or stew recipes.

One-Pot Pork and Vegetable Bake

Makes 4 to 6 servings

This slow-cooked dish is perfect for when the weather keeps you indoors and a warming stew is welcome at the end of the day. You can serve it with couscous, rice or slices of whole wheat bread.

- 2 tbsp Vietnamese Five-Spice Blend (page 143)
- 2 tbsp coconut sugar crystals
- 2 tbsp coconut flour
- 2 lb (1 kg) boneless pork loin, cut into 1-inch slices
- 3 tbsp melted coconut oil, divided
- 1 onion, cut into quarters
- 2 apples, cut into quarters
- 1 sweet potato, cut into 1-inch cubes
- 2 cups chopped kale or spinach, fresh or frozen
- 1 can (28 oz/796 mL) diced tomatoes and juices
- 1/4 cup freshly squeezed lemon juice

1. Preheat oven to 375°F.
2. In a shallow bowl or pie plate, combine spice blend, sugar crystals and flour. Dredge pork slices through the flour mixture to evenly coat both sides. Transfer to a plate and set aside. Reserve remaining flour mixture.
3. In a Dutch oven, heat 2 tbsp oil over medium heat. Add pork slices and cook, turning frequently, for 2 to 3 minutes or until browned on both sides. Transfer to a plate and set aside.
4. Add remaining oil, onion, apples, sweet potato and reserved flour mixture (if any) and cook, stirring constantly, for 2 minutes or until lightly browned and well coated. Add kale and stir to combine. Remove pot from heat. Transfer half of the apple mixture to a bowl. Layer the reserved pork over the remaining vegetables in the pot and top with the reserved apple mixture. Set aside.
5. In a bowl, combine tomatoes and their juices with lemon juice. Pour over the layered meat and vegetables. Cover and bake in preheated oven for 1 hour or until vegetables are tender and meat reaches 160°F on a meat thermometer.

My Vietnamese Five-Spice Spice Blend offers a twist on the more common Chinese five-spice blend, which you can substitute in this recipe.

Chopped, frozen kale (shown here) or spinach are excellent green vegetables to have on hand for dishes like this one.

A Dutch oven is a large cooking pot with a lid used to simmer foods on top of the stove or in the oven for long periods of time. It is traditionally made of cast iron or ceramic-coated cast iron, but stainless steel versions are available. They are heavy, and the lids are usually dimpled in order to condense steam and return the liquids to the pot. If you don't have a Dutch oven, you can use a large cast iron skillet or a pot, with a lid, that is both flame- and oven-proof.

Seasonal Vegetable Biryani

Makes 4 servings

Now that people are starting to understand the cancer-fighting properties of fruits and vegetables, great vegetarian entrées are more in demand than ever. This recipe offers everything you could want in a main dish: protein (from the lentils), fantastic flavour (from the spices, the tart cranberries and the coconut) and versatility (you can use whatever is convenient and on-hand).

- 3/4 cup dried lentils (red, yellow or green)
- 2 tbsp melted coconut oil
- 1 onion, chopped
- 2 cloves garlic, coarsely chopped
- 1 tbsp Garam Masala Spice Blend (page 137)
- 2 tsp Madras Curry Spice Blend (page 139)
- 1 tsp sea salt
- 1 cup basmati rice, rinsed and drained
- 2-1/2 cups vegetable or chicken broth
- 2 cups chopped mixed vegetables
- 1 cup green peas, fresh or frozen
- 1/4 cup dried cranberries or raisins
- 1/2 cup toasted shredded coconut, fresh or dried (optional)

1. In a saucepan, bring 2 cups water to a boil. Add lentils, partially cover, reduce heat to low and simmer for 30 to 40 minutes or until tender. Drain and set aside.
2. Meanwhile, in the bottom of a large flameproof tagine or Dutch oven, heat oil over medium heat. Sauté onion for 5 minutes. Add garlic, garam masala blend, curry spice blend and salt and cook, stirring frequently, for 1 to 2 minutes or until garlic is softened but not brown.
3. Add rice and stir well to coat with spices. Add broth and bring to a boil. Stir in vegetables, cover, reduce heat to low and simmer for about 20 minutes or until rice is tender.
4. Add peas and cook for 3 to 4 minutes or until heated through. Stir in cranberries and cooked lentils. To serve, sprinkle with toasted coconut if desired.

Fresh seasonal vegetables—asparagus, carrots, parsnips, squash, zucchini, cabbage—are perfect for this recipe, but if it is more convenient, use 2 cups diced or sliced frozen mixed vegetables.

To toast coconut: Preheat oven to 350°F. Arrange coconut in an even layer on a rimmed baking sheet and toast, stirring once or twice, in preheated oven for 5 to 7 minutes or until lightly brown.

Shrimp in Coconut Sauce

Makes 4 servings

This coconut–garam masala–almond combination transforms the shrimp into a divine dish reminiscent of faraway Thailand. It is spicy and yet sweet, and the texture of the sauce is thick and creamy without dairy fat. Although this dish is guest-worthy, don't wait for special occasions to serve it.

- 1 tbsp Garam Masala Spice Blend (page 137)
- 1 tbsp coconut flour
- 1 tsp sea salt
- 4 tbsp melted coconut oil
- 1 onion, chopped
- 1 leek, white and light green parts, chopped
- 3 cloves garlic, coarsely chopped
- 1/2 cup slivered almonds
- 1 red bell pepper, chopped
- 1/2 head cauliflower, cut into florets
- 1 can (14 oz/398 mL) coconut milk, divided
- 1 lb (500 g) shelled and deveined shrimp
- 1 cup cooked wheat berries or spelt

1. In a small bowl, combine spice blend, flour and salt. Set aside.
2. In a Dutch oven or the bottom of a large tagine, heat oil over high heat. Sauté onion and leek for 3 to 5 minutes or until translucent. Add garlic and almonds and cook, stirring frequently, for 1 minute. Add spice mixture and cook, stirring frequently, for 1 minute.
3. Add red pepper, cauliflower and 1 cup of the milk. Cover, reduce heat to low and cook for 6 to 10 minutes or until vegetables are tender. Add remaining milk and bring to a boil over medium-high heat. Reduce heat to medium-low and add shrimp and cook, stirring frequently, until the shrimp turn pink. Add cooked wheat berries and stir until heated through.

You can substitute the cooked wheat berries with an equal amount of cooked rice or quinoa.

For extra flavour, try adding 1/4 cup chopped mixed fresh herbs (parsley, oregano or thyme) or Coconut Gremolata (page 149) at the very end of Step 3.

Spiced Slow-Cooked Chicken

Makes 4 to 6 servings

This recipe is a delicious, easy weekday meal fix and it is just as good, or even better, the next day for lunch. (You'll need a slow cooker to prepare this dish.)

- 2 tbsp melted coconut oil
- 2 onions, cut into quarters
- 1 red bell pepper, chopped
- 4 cloves garlic, finely chopped
- 2 tbsp creamed coconut
- 1 cup coconut water or plain water
- 1 tbsp Garam Masala Spice Blend (page 137)
- 1 tsp ground cumin
- 1 tsp ground turmeric
- 2 lb (1 kg) skinless chicken thighs
- 1 can (28 oz/796 mL) whole or diced tomatoes with juices
- 1 can (19 oz/540 mL) chickpeas, drained and rinsed

1. In a skillet, heat oil over medium heat. Sauté onions and red pepper for 5 minutes or until soft. Add garlic, creamed coconut, coconut water, spice blend, cumin and turmeric and cook, stirring constantly, for 2 minutes. Remove from heat and set aside.
2. In the bowl of a slow cooker, combine chicken, tomatoes and juices, and chickpeas. Add reserved vegetables and spices and mix well. Cover and cook on low for 3 to 4 hours or on high for 1-1/2 to 2 hours, until chicken is cooked through and easily pulls apart.

If you don't have creamed coconut, substitute 1 cup of vegetable or chicken broth for the creamed coconut and water.

If desired, you can substitute 4 to 6 boneless skinless chicken breasts for the chicken thighs.

Sweet Potato Shrimp Cakes

Makes 8 to 10 3-inch cakes

Make these easy seafood cakes the night before for a company-is-coming appetizer or main dish, or just for a delicious change. Robust, sweet coconut is the perfect match for seafood—you can use lobster, crab or scallops—and the tartly sweet pineapple sauce completes the dish.

- 2 medium sweet potatoes, cooked
- 1 cup cooked shrimp, shelled and deveined, finely chopped
- 2 tbsp chopped fresh basil
- 2 tbsp cocktail sauce or tomato sauce
- 1 large egg, beaten
- 1 can (5.4 oz/160 mL) coconut cream
- 2 tbsp unsweetened shredded coconut, fresh or dried
- 1 cup coconut flour or as needed
- 1/2 cup melted coconut oil or as needed
- 2 cups Sweet and Sour Pineapple Sauce (page 154)

1. Preheat oven to 300°F.
2. In a bowl, mash sweet potatoes. Add shrimp, basil, cocktail sauce, egg, cream and shredded coconut. Using your hands, work the mixture into a smooth batter. Sprinkle flour over the batter and knead into a soft dough, adding enough flour to make the dough thick enough to form into cakes. Using your hands, pat the dough into 3-inch-round cakes, about 1-inch thick. Lightly dust both sides of the cakes in flour and transfer them to a plate.
3. In a skillet, heat oil over medium heat. Add 3 to 4 cakes to the skillet and fry for 3 minutes on each side or until a golden crust forms on both sides. Transfer to a baking sheet and keep warm in preheated oven. Repeat with remaining cakes, adding more oil as required. Serve with pineapple sauce.

Bahmi (Pork Noodles)

Makes 4 servings

I've used the Polynesian Spice Blend here because its subtle sweetness complements the pork. For a hotter, more complex taste, punch it up a notch with the same amount of Hot Spice Paste (page 138) or Madras Curry Spice Blend (page 139). Either way, this makes a fast, delicious meal.

- 1/2 lb (250 g) noodles
- 2 tbsp Polynesian Spice Blend (page 140)
- 2 tbsp coconut flour
- 1/2 lb (250 g) boneless pork loin, cut into 1/2-inch cubes
- 2 tbsp melted coconut oil
- 2 onions, chopped
- 1 carrot, finely chopped
- 4 cloves garlic, minced
- 1 cup coconut water or coconut milk
- 3 tbsp coconut aminos or tamari sauce
- 1 small head bok choy, coarsely shredded

1. In a pot of salted boiling water, cook noodles according to package instructions. Drain and rinse under cool running water. Set aside.
2. In a bowl, combine spice blend and flour. Add pork cubes to spice mixture and toss to coat well.
3. In a wok or skillet, heat oil over medium heat. Sauté onions for 5 minutes. Add carrot, garlic and pork, scraping any loose seasoning into the pan. Cook, stirring frequently, for 4 to 5 minutes or until pork is browned all over. Add coconut water and aminos and bring to a boil. Cover, reduce heat and simmer for 8 to 10 minutes or until vegetables are tender and pork is cooked through and reaches 160°F on a meat thermometer.
4. Add bok choy and cooked noodles and simmer, stirring frequently, for 2 minutes or until bok choy is wilted and noodles are heated through.

Use egg noodles, vermicelli, spaghettini, linguini, fettuccine, rice or soba noodles.

My Polynesian Spice Blend is a sweet mixture that includes cinnamon and other sweet spices. Garam Masala Spice Blend may be substituted.

VARIATION: For a vegetarian version, substitute 2 cups cubed eggplant or 1 cup cubed tempeh for the pork. Reduce the cooking time in Step 3 and cook until eggplant is tender.

Coconut Rice

Makes 4 to 6 servings

Sweet coconut marries mild, slightly nutty rice and they both live together happily ever after. It really is a match made in heaven. Even though this recipe calls for long-grain rice, I have used all kinds of rice in this dish—short-grain brown, basmati and even China's forbidden black rice. (Increase the cooking time by about 10 minutes when you use denser, whole-grain rice.)

- 2 tbsp melted coconut oil
- 1 onion, finely chopped
- 3 tbsp unsweetened coconut flakes, fresh or dried
- 2 cups long-grain rice
- 2 cans (14 oz/398 mL each) coconut milk

1. In a saucepan, heat oil over medium heat. Sauté onion for 5 minutes. Add coconut flakes and rice and cook, stirring constantly, for 1 minute or until rice is translucent and resembles glass.
2. Add milk and bring to a boil. Cover, reduce heat and simmer for 30 minutes or until the liquid has been completely absorbed and the rice is tender. Remove from heat and set aside for 3 minutes. Fluff with a fork and serve.

Bulgur and Lentils with Coconut Gremolata

Makes 4 servings

I keep coconut water on hand to use whenever a light, vegan broth would be more appropriate than beef or chicken broth.

- 3/4 cup dried lentils
- 2 tbsp melted coconut oil
- 1 onion, finely chopped
- 1/2 red bell pepper, finely chopped
- 2 cloves garlic, minced
- 1 tbsp Polynesian Spice Blend (page 140) or Garam Masala Spice Blend (page 137)
- 1 cup bulgur
- 2 cups coconut water or plain water
- 1 tbsp grated lemon rind
- 2 tbsp freshly squeezed lemon juice
- Sea salt and freshly ground black pepper
- 1/4 to 1/2 cup Coconut Gremolata (page 149), for garnish (optional)

1. In a saucepan, cover lentils with hot water. Bring to a boil over high heat. Reduce heat and simmer for 20 minutes or until lentils are tender but firm. Drain and rinse under cool running water. Set aside.
2. In a wok or Dutch oven, heat oil over medium heat. Sauté onion and red pepper for 5 minutes. Add garlic, spice blend and bulgur and cook, stirring constantly, for 1 minute. Add coconut water and bring to a boil. Add lentils, reduce heat and simmer for 5 minutes. Remove from heat and set aside for 15 minutes or until bulgur is tender and all of the liquid has been absorbed.
3. Stir in lemon rind and juice. Season to taste with salt and pepper. Serve topped with 1 tablespoon of gremolata per serving or pass separately at the table (if using).

Lentils (*Lens culinaris*) are the edible seeds of plants in the legume family. You can distinguish lentils from other dried legumes because they are small, flat and round, with 2 disc-like halves per seed. Referred to as "pulses," lentils grow in pods and are dried for easy storage and transportation, so they are often confused with dried beans and peas. Unlike dried peas and beans, lentils do not require pre-soaking and are tender within about 20 minutes when simmered in water. All legumes are high in carbohydrates, protein and fibre, which makes them an essential ingredient in meat-free dishes.

Bulgur is a hard whole wheat grain that has been parboiled, dried and cracked, which makes it fast and easy to cook. Bulgur varies in type, from fine or "instant" to medium and coarse. It has a slightly nutty flavour and a tender chewy texture and is often used in place of couscous in Middle Eastern dishes, especially tabouli salad. Bulgur can also be substituted for other whole grains, such as barley or quinoa, in most recipes.

Coconut Barley Risotto

Makes 4 servings

I like this recipe because, unlike most risotto methods, it doesn't require you to stand over the pan for almost an hour stirring in small amounts of broth. Instead, you let the barley cook in a covered pan and occasionally check that it has enough liquid to keep cooking until tender.

- 6 cups chicken broth, divided
- 2 boneless skinless chicken breasts
- 2 tbsp melted coconut oil
- 1 tbsp butter
- 1 onion, finely chopped
- 2 cloves garlic, minced
- 2 cups sliced shiitake or cremini mushrooms
- 1 cup pearl barley
- 1/2 cup dry white wine
- 3 cups chopped spinach or kale, fresh or frozen
- 1 cup unsweetened coconut flakes, fresh or dried
- 1/2 cup freshly grated Parmesan cheese
- Sea salt and freshly ground black pepper

1. In a saucepan, bring broth to a boil over high heat. Add chicken, cover, reduce heat and simmer for 8 minutes or until chicken is cooked through and reaches 165°F on a meat thermometer. Transfer chicken to a plate or cutting board and set aside to cool. Cover broth and keep warm on the element. Once chicken is cool, shred or dice and set aside.
2. Meanwhile, in a saucepan, heat oil and butter over medium heat. Add onion and sauté for 5 minutes. Add garlic and mushrooms and cook, stirring frequently, for 5 minutes or until mushrooms release their juices and begin to brown. Add barley and cook, stirring constantly, for 1 minute.
3. Add wine, bring to a boil and boil for 1 minute. Add 5 cups broth and return to a boil. Cover, reduce heat and simmer, stirring often, for 45 minutes or until barley is tender. If barley becomes dry during cooking, add 1/4 cup more broth at a time, as necessary.
4. Stir in spinach and coconut flakes and, if necessary, add more broth to keep the barley moist. Cook, uncovered, over medium heat for 2 minutes or until spinach is wilted. Stir in reserved chicken and more broth, if required, to keep the barley glistening with moisture but not swimming in broth. Cook, stirring constantly, to warm through. Stir in cheese and season to taste with salt and pepper.

Risotto is a creamy Italian short-grain rice dish that is slow-cooked to a tender and silky texture with a rich, thick, smooth consistency. At its most basic, it is cooked by adding small amounts of hot broth to the pan over time until the rice is tender. By slowly adding the broth and stirring the rice until it is absorbed before adding more broth, you allow the rice to take up much more liquid than it would if it were simply simmered in the broth. In addition, the longer, slow cooking allows the starch to thicken the liquid for a rich sauce. A hard Italian cheese is grated into the hot rice toward the end of the cooking. All kinds of other grains may be used in risotto-style dishes, and a variety of other ingredients may be added, including mushrooms, asparagus, squash, cauliflower, fresh, wild or frozen greens or nuts.

Barley (*Hordeum vulgare*), a cereal grain, may be substituted for rice in many recipes. As all grains, it is available with only the tough outer hull removed (pot or Scotch barley) or polished to remove the bran layer (pearl barley). If available, use the darker pot or Scotch barley, but even pearl barley, with its significant amount of fibre distributed throughout the kernel, is a healthy choice over other processed grains. Barley is not suitable for gluten-free diets.

Farro Stir-Fry

Makes 4 servings

To save time, I often use canned red peppers and canned artichoke hearts in this dish, but you can easily adapt this recipe by using chopped fresh seasonal vegetables: peas and sweet baby carrots in summer, root vegetables in winter.

- 1 cup farro, rinsed and drained
- 1 tbsp melted coconut oil
- 2 leeks, white and light green parts, sliced
- 1 tbsp Polynesian Spice Blend (page 140) or Garam Masala Spice Blend (page 137)
- 2 cups chopped kale or spinach, fresh or frozen
- 1/2 cup chopped roasted red peppers
- 1 can (14 oz/398 mL) artichoke hearts, drained and cut into quarters
- 2 tbsp coconut aminos or tamari sauce
- Sea salt and freshly ground black pepper

1. In a saucepan, combine farro with 2-1/2 cups water and 1/2 tsp salt. Bring to a boil over high heat. Cover, reduce heat and simmer for 25 to 30 minutes or until the liquid has been absorbed and the farro is tender. Remove from heat and set aside for 3 minutes. Fluff with a fork.
2. In a wok or skillet, heat oil over medium heat. Add leeks and spice blend and cook, stirring frequently, for 4 minutes or until tender-crisp. Add cooked farro, kale, red peppers, artichoke hearts and aminos and cook, stirring constantly, for 2 minutes or until the leeks are tender and the farro and vegetables are heated through. Season to taste with salt and pepper.

Farro (*Triticum dicoccon*) is Italian for an ancient form of wheat known as "emmer." More popular in Europe than North America, farro is a nutty-tasting, slightly sweet, chewy whole grain. If desired, you can substitute spelt or regular wheat berries in recipes that call for farro. Farro is not suitable for gluten-free diets.

Herbed Rice Pilaf

Makes 4 servings

I serve this elegant rice dish with stir-fried entrées and saucy main dishes such as Spiced Slow-Cooked Chicken (page 89) or Shrimp in Coconut Sauce (page 88). It teams up nicely with baked chicken or fish, and because of the vegetables, it completes a meal.

- 2 tbsp melted coconut oil
- 3 green onions, white and green parts, chopped
- 1 clove garlic, finely chopped
- 1 cup long-grain white or brown rice
- 2 cups chicken broth
- 1 cup chopped spinach, fresh or frozen
- 1/2 cup chopped fresh parsley
- 2 tsp chopped fresh rosemary

1. In a saucepan, heat oil over medium heat. Sauté onions for 3 minutes. Add garlic and rice and cook, stirring constantly, for 1 minute or until rice is translucent and resembles glass.
2. Add broth and bring to a boil. Cover, reduce heat and simmer for 20 minutes. Stir in spinach, parsley and rosemary, cover and simmer for 10 minutes or until the liquid has been absorbed and rice is tender. Remove from heat and set aside for 3 minutes. Fluff with a fork.

If you are using brown rice, increase the cooking time to 30 minutes before adding the spinach in Step 2.

Nasi Goreng (Indonesian Fried Rice)

Makes 4 servings

This recipe is much easier to make if you cook the chicken and rice the night before. It is essential that the rice be chilled before adding to the stir-fry—if not, it will get sticky and clump together instead of being coated with the coconut oil and spice blend.

- 4-1/2 cups chicken broth
- 1 boneless skinless chicken breast
- 2 cups short-grain brown rice
- 3 tbsp melted coconut oil
- 1 onion, finely chopped
- 1 red or green bell pepper, chopped
- 1 cup fresh bean sprouts
- 2 cloves garlic, finely chopped, or 2 tbsp Coconut Gremolata (page 149)
- 1 tbsp Polynesian Spice Blend (page 140) or Coconut Jerk Spice Rub (page 134)
- 1/2 lb (500 g) small shrimp, peeled and deveined
- 2 tbsp coconut aminos or tamari sauce
- Sea salt and freshly ground black pepper

1. In a saucepan, bring broth to a boil over high heat. Add chicken, cover, reduce heat and simmer for 8 minutes or until chicken is cooked through and reaches 165°F on a meat thermometer. Lift out chicken and place on a plate or cutting board (reserving pan with broth); set aside until cool enough to handle. Shred or dice chicken, cover tightly and refrigerate until ready to use.
2. Stir rice into hot broth. Cover and simmer for 40 minutes or until the liquid has been completely absorbed and the rice is chewy-tender. Set aside to cool and then chill for at least 30 minutes or overnight. (Recipe may be made ahead up to this point.)
3. In a wok or skillet, heat oil over medium heat. Sauté onion for 5 minutes. Add bell pepper and cook, stirring constantly, for 5 minutes or until vegetables are tender. Add cooked, chilled rice, shredded chicken, bean sprouts, garlic, spice blend, shrimp and aminos and cook, stirring constantly, for 2 minutes or until shrimp are cooked (turn pink) and chicken and rice are heated through. Season with salt and pepper to taste.

For an authentic fried rice dish, garnish it with a thin omelette chopped into strips: After transferring the finished rice to a warm platter, crack 2 eggs into the hot wok and cook over medium-low heat, lifting the egg while cooking and allowing any liquid to run underneath. Transfer the finished omelette onto a plate and cut into strips.

Roasted Cauliflower Rigatoni

Makes 6 servings

Make this a vegetarian main dish by adding 2 cups of cooked chickpeas or lentils and a 1/2 cup of chopped nuts to the roasted cauliflower in Step 3. Make it gluten-free by using rice pasta and gluten-free breadcrumbs. Make it any way you like, but make it often because it could become a family favourite.

- 1 large head cauliflower, trimmed and cored
- 2 onions, cut into eighths
- 1 head garlic, cloves peeled
- 1 tbsp fresh thyme leaves
- 3 tbsp melted coconut oil
- 1 tsp sea salt
- 3/4 lb (375 g) rigatoni or penne pasta
- 1 cup coconut milk
- 1/2 cup breadcrumbs
- 3/4 cup shredded cheddar cheese

1. Preheat oven to 300°F. Line a rimmed baking sheet with parchment paper. Lightly butter a 9- × 13-inch casserole dish.
2. Using a sharp knife, quarter the cauliflower. Cut each quarter into 1/4-inch-thick slices and transfer to a bowl. Add onions, garlic and thyme and mix to combine. Add oil and toss to coat. Roast in preheated oven for about 20 minutes, stirring halfway through, until cauliflower is browned on edges and tender. Sprinkle with salt and toss to mix. Do not turn oven off.
3. Meanwhile, cook the pasta in a large pot of boiling salted water according to package instructions. Drain and transfer to a bowl. Add roasted cauliflower and onion mixture and milk. Stir well and transfer to the prepared casserole dish.
4. Spread breadcrumbs evenly overtop, followed by cheese. Bake for 15 minutes or until the breadcrumbs are browned and the cheese is melted.

The best way to coat the cauliflower and onion with the oil in Step 2 is to use your hands; if the vegetables are cool, the oil will solidify. Don't worry if it does—it will melt in the oven and be evenly distributed when you stir the vegetables halfway through cooking them.

Reserve the bowl that you use to toss the cauliflower and onions; you can use it to toss together the roasted vegetables and the cooked rigatoni.

Soba Noodles with Peanut Sauce

Makes 4 servings

I love options and I try to build them into my recipes so that they appeal to cooks, kids and everyone in between. Experiment with this dish by using different kinds of vegetables or one of the tasty sauces in Chapter 6.

- 8 oz soba noodles
- 2 carrots, cut into 1-inch pieces
- 1/2 lb (250 g) green beans, trimmed and cut into 1-inch pieces
- 2 cups coarsely chopped kale or spinach, fresh or frozen
- 1 cup fresh bean sprouts or green peas
- 2 cups Shiitake Peanut Sauce (page 159) or Peanut Papaya Sauce (page 160)
- 4 green onions, white and green parts, sliced into 1/4-inch pieces

1. Bring a large pot of salted water to a boil over high heat. Cook noodles according to package instructions. Using tongs, transfer noodles to a colander. Rinse under cool running water and set aside. Keep water boiling in pot.
2. Add carrots to pot and cook for 1 minute. Add beans and cook for 1 minute. Add kale and cook for 1 minute. Drain vegetables and transfer to a large bowl. Add cooked noodles, bean sprouts and peanut sauce and toss to mix well. Serve warm or at room temperature.

Baked Spinach

Makes 4 servings

It is fascinating that so many cultures have found a way to combine green leaves with cream. The English make a white sauce with butter, onion, flour and cream for their stovetop creamed spinach. The French version, *épinards à la crème*, is a simple combination of double cream and butter with spinach.

Half a world away, across all of Oceania, at every occasion and every table—from humble to grand—the young leaves of the taro plant are either first wrapped in banana leaves or simply mixed with coconut milk and baked. This dish substitutes spinach (or kale) for the taro leaves, but it still has a tropical flavour.

1 can (14 oz/398 mL) coconut milk
1 tbsp tahini
1/2 tsp ground nutmeg
1/2 tsp sea salt
1 lb (500 g) spinach (or kale), steamed and chopped
1/4 cup toasted coconut flakes, fresh or dried

1. Preheat oven to 325°F. Lightly oil an 11- × 7-inch casserole dish.
2. In casserole dish, combine milk, tahini, nutmeg and salt. Add spinach and toss to mix well. Bake in preheated oven for 20 minutes or until spinach is tender. Sprinkle toasted coconut evenly overtop.

Chopped frozen spinach (or kale) is an excellent green vegetable to have on hand for one-pot dishes like this one.

To toast coconut: Preheat oven to 350°F. Arrange coconut in an even layer on a rimmed baking sheet and toast, stirring once or twice, in preheated oven for 5 to 7 minutes or until lightly brown.

Broccoli with Sesame Dressing

Makes 4 servings

Easy and delicious, this stir-fry can be enjoyed hot as a side dish or served at room temperature as a salad. The Sesame Dressing adds a nuttiness to the overall dish that both tames and complements the broccoli.

- 2 tbsp melted coconut oil
- 1-1/2 lb (750 g) broccoli, cut into 2-inch spears
- 1 onion, cut into quarters
- 1 cup Sesame Dressing (page 153)
- 2 tbsp sesame seeds, for garnish (optional)

1. In a skillet or wok, heat oil over medium heat. Add broccoli and onion and cook, stirring frequently, for 1 minute. Add sesame dressing, stir well and bring to a simmer. Cook, stirring frequently, for 3 to 5 minutes or until broccoli is tender-crisp. Garnish with sesame seeds (if using).

Cashew Vegetable Ragout

Makes 4 servings

For a hearty vegetarian meal, serve this ragout with Coconut Polenta (page 111) or any of the rice recipes in this chapter.

- 2 tbsp melted coconut oil
- 1 onion, chopped
- 2 cloves garlic, finely chopped
- 1 tbsp fresh thyme leaves
- 2 carrots, cut into 1-inch pieces
- 1 leek, white and green parts, sliced
- 1 sweet potato, cut into 1-inch cubes
- 1 can (28 oz/796 mL) diced tomatoes and juices
- 1 can (5.4 oz/160 mL) coconut cream
- 1 cup coarsely chopped spinach, fresh or frozen
- 1/4 cup cashew butter
- 1/4 cup roasted salted cashew pieces

1. In a wok or skillet, heat oil over medium heat. Sauté onion for 5 minutes. Add garlic and thyme and cook, stirring constantly, for 1 minute. Add carrots, leek and sweet potato and cook, stirring constantly, for 30 seconds or until vegetables are well mixed.
2. Add tomatoes and juices and bring to a boil. Reduce heat and simmer for 15 minutes or until vegetables are tender. Add cream and spinach and cook for 1 minute or until spinach is wilted. Stir in cashew butter and cashews and heat through.

If cashew butter is not available, you can substitute an equal amount of peanut butter.

Chard and Chickpeas

Makes 4 servings

Easy-peasy, as they say—so easy I tend to make this often. Sometimes if I'm home alone at dinner, this is all I prepare because it's quite filling. Blanching greens and cruciferous vegetables (such as spinach, kale, broccoli, cauliflower and Brussels sprouts) destroys the goitrogens that, when eaten raw, can put stress on the thyroid, so don't skip Step 1.

- 1-1/2 lb (750 g) Swiss chard
- 2 tbsp melted coconut oil
- 1 onion, chopped
- 4 cloves garlic, finely chopped
- 1/4 cup chopped pistachio nuts
- 1 can (19 oz/540 mL) chickpeas, rinsed and drained
- 1/4 cup freshly squeezed lemon juice

1. In a pot of salted boiling water, blanch Swiss chard for 4 to 5 minutes or until tender. Drain and rinse under cold running water. Pat dry, coarsely chop and set aside.
2. In a skillet, heat oil over medium heat. Sauté onion for 5 minutes. Add garlic and nuts and cook, stirring constantly, for 1 minute. Turn heat off and add cooked Swiss chard, chickpeas and lemon juice. Toss well to combine.

Cheesy Baked Eggplant

Makes 4 servings

Toss and bake, use one dish and relax. What could be easier? This is a great vegetable side dish that when teamed up with a grain dish makes a complete light meal (like many of the recipes in this section).

- 1 eggplant, cut into 1-inch cubes
- 1 onion, cut into quarters
- 1 cup Honey Lime Coconut Dressing (page 146)
- 1 cup grated Asiago cheese

1. Preheat oven to 375°F. Lightly oil an 11- × 7-inch casserole dish.
2. In prepared casserole dish, toss eggplant, onion and dressing. Bake in preheated oven for 45 minutes. Stir and sprinkle with cheese. Bake for 15 minutes or until eggplant is tender and cheese is bubbly.

Coconut Polenta

Makes 12 squares

I enjoy the crisp texture of grilled polenta with stir-fried seasonal vegetables, especially the Green Shiitake Stir-Fry (page 113) and the Seasonal Vegetable Biryani (page 87). If you are in a hurry, you can save time and omit the grilling in Step 4.

- 1 cup cornmeal
- 1 cup corn kernels, fresh or frozen
- 1/2 cup coconut chips or flakes, fresh or dried
- 1/2 cup grated Asiago cheese

1. Lightly oil a 9- × 13-inch baking dish.
2. In a saucepan, bring 4 cups of salted water to a boil over high heat. Reduce heat to low and gradually whisk in cornmeal. Cook, stirring often, for 10 minutes. Add corn and coconut and cook, stirring often, for 5 to 10 minutes or until polenta is thick enough to mound on a spoon. Remove from heat and stir in cheese.
3. Spread polenta in prepared baking dish. Put aside to allow the polenta to set, about 30 minutes.
4. Using a sharp knife, cut into 12 squares. Grill on oiled grill pan over medium-high heat, turning once, for 5 minutes or until crispy and hot.

Green Shiitake Stir-Fry

Makes 4 servings

If food is indeed medicine, as I believe, this dish tops the list of superfoods. Shiitake mushrooms have incredible cancer-fighting components, and the garlic and coconut oil simply compound their amazing properties.

- 3 tbsp melted coconut oil
- 2 cloves garlic, finely chopped
- 1 tbsp finely chopped candied ginger or gingerroot
- 1/2 lb (250 g) fresh shiitake mushrooms, stemmed and thinly sliced
- 1 lb (500 g) fresh asparagus or green beans, cut diagonally into 2-inch pieces
- 1/3 cup chicken broth
- 2 tbsp coconut aminos or tamari sauce
- 1 tbsp coconut nectar or honey
- 2 cups baby spinach leaves
- 1 sheet toasted nori, thinly sliced (optional)

1. In a wok or skillet, heat oil over medium heat. Add garlic and ginger and cook, stirring constantly, for 1 minute or until fragrant but not browned. Add mushrooms and cook, stirring frequently, for 2 minutes or until beginning to brown. Add asparagus and cook, stirring constantly, for 3 minutes or until tender-crisp.
2. Add broth, aminos and nectar and bring to a boil. Add spinach and cook, stirring frequently, for 30 seconds or until spinach wilts. Using a slotted spoon, transfer vegetables to a warmed serving dish and set aside.
3. Boil liquid in wok for 3 to 5 minutes or until reduced and slightly thickened. Pour over vegetables and garnish with nori (if using).

Nori (*Poryphyra*) is the Japanese name for an edible seaweed that is pressed into very thin sheets and dried. It is most often used for rolling sushi. Nori sheets may be purchased from most well-stocked grocery stores and specialty food stores.

To toast nori: Using tongs, hold individual sheets over a flame or very hot element for about 30 seconds or until crisp or toasted. You can use kitchen scissors to cut toasted nori sheets into very thin strips for garnishing vegetable or pasta dishes.

Island Vegetable Sauté

Makes 4 servings

Let this dish whisk you away to a private cove in the tropics. Don't be put off by the long list of ingredients: once you trim and chop the vegetables, it takes less than 15 minutes to stir-fry them into a perfect accompaniment to any of the main dishes in this chapter. So chop and stir and dream a little as the essence of pure coconut wafts lightly around you.

- 2 tbsp melted coconut oil
- 1 onion, chopped
- 2 cloves garlic, finely chopped
- 1 tbsp Polynesian Spice Blend (page 140) or Garam Masala Spice Blend (page 137)
- 1 eggplant, diced
- 1 green apple, peeled and diced
- 1 mango, diced
- 1 small zucchini, diced
- 1 cup green or yellow beans, cut into 1-inch pieces
- 2 large tomatoes, coarsely chopped
- 2 tbsp freshly squeezed lemon juice
- 1 tbsp coconut aminos or tamari sauce
- 1/4 cup grated fresh coconut or shredded unsweetened fresh or dried coconut

1. In a wok or skillet, heat oil over medium heat. Sauté onion for 5 minutes. Add garlic and spice blend and cook, stirring constantly, for 1 minute. Add eggplant, apple, mango, zucchini and beans and cook, stirring constantly, for 5 minutes or until vegetables and fruit are tender-crisp.
2. Add tomatoes, lemon juice and aminos and cook, stirring frequently, for 3 minutes or until tomatoes have released their juices. Stir in grated coconut and heat through.

It's best to use freshly grated coconut in this recipe, but if you don't have fresh coconut on hand, soak dried shredded unsweetened coconut in 1/4 cup water while you prepare the vegetables. Drain (saving the soaking water for other recipes) and add reconstituted coconut in Step 2.

Three Bean Stir-Fry

Makes 4 servings

Bean sprouts are a simple but superior nutrition source because they contain the elements a person needs to grow and be healthy: enzymes, protein, antioxidants, vitamins and minerals. The process of sprouting increases the amount of B vitamins (including folate), vitamin C, fibre and essential amino acids in the grains.

- 2 tbsp melted coconut oil
- 2 leeks, white and light green parts, coarsely chopped
- 1/4 tsp red pepper flakes
- 1 cup sliced green beans
- 1 cup sliced yellow beans
- 2 cups fresh bean sprouts
- 1/4 cup toasted coconut flakes, fresh or dried
- 2 tbsp roasted salted peanuts or cashews, crushed

1. In a wok or skillet, heat coconut oil over medium heat. Sauté leeks and red pepper flakes for 5 minutes. Add green and yellow beans and cook, stirring constantly, for 5 minutes or until vegetables are tender-crisp. Stir in bean sprouts and cook, stirring frequently, for 1 minute or until heated through. Stir in coconut flakes and peanuts and heat through.

Fresh long beans (*Vigna unguiculata*, subspecies *sesquipedalis*), often called Chinese long beans, are delicious in this dish. You can find them at well-stocked grocery stores and Asian markets. Trim and slice them into 1-inch pieces as you would regular green or yellow beans.

Sprouts are easy to digest, and they add a bit of crunch to stir-fry dishes like this one. For variety, look for interesting, different-coloured sprouts such as broccoli, sunflower or spelt sprouts.

Potato-Stuffed Acorn Squash

Makes 4 servings

This is a pretty substantial side dish. If you add a cup or two of cooked lentils or legumes (dried peas or beans) for protein, it makes a great entrée and a beautiful addition to the buffet table at Thanksgiving and Christmas.

- 2 acorn squash
- 2 tbsp melted coconut oil
- 1 small onion, chopped
- 2 cloves garlic, finely chopped
- 1 tbsp Garam Masala Spice Blend (page 137)
- 3 cups chopped fresh spinach
- 1 cup grated sweet potato
- 1 cup cooked chickpeas, rinsed and drained
- 1/4 cup chopped macadamia nuts or almonds
- 3 tbsp unsweetened coconut flakes, fresh or dried

1. Preheat oven to 350°F. Line a rimmed baking sheet with parchment paper.
2. Cut squash in half lengthwise. Scoop out and discard seeds. Arrange on prepared baking sheet, cut sides up.
3. In a skillet, heat oil. Sauté onion for 5 minutes. Add garlic and spice blend and cook, stirring frequently, for 1 minute. Transfer to a bowl.
4. Add spinach, potato, chickpeas, nuts and coconut flakes and mix well. Spoon equal amounts into squash. Bake in preheated oven for 45 minutes or until squash is tender-crisp.

5

Appetizers & Snacks

Apple Walnut Salsa

Makes 1-1/2 cups

I'm not a big fan of cilantro. The truth is, I detest it. So I'm always making unique salsas that are devoid of the only herb I can't stand. This is my new favourite, and it's certainly unique: it's cooked when most are raw; it contains apples and walnuts; and the sweet and sour essence of the coconut vinegar and sugar boost the flavours into a new stratosphere. I use it to garnish all sorts of dishes, including Bean and Nut Loaf (page 73) and One-Pot Pork and Vegetable Bake (page 84). You can add as much chopped fresh cilantro as you like!

- 2 tbsp melted coconut oil
- 1 onion, coarsely chopped
- 1 orange or red bell pepper, coarsely chopped
- 4 cloves garlic, finely chopped
- 2 apples, peeled and coarsely chopped
- 1 tbsp Madras Curry Spice Blend (page 139)
- 1/4 cup coconut vinegar
- 1/4 cup coconut sugar crystals
- 1 tsp sea salt
- 1/2 cup coarsely chopped walnuts

1. In a saucepan, heat oil over medium heat. Sauté onion and bell pepper for 5 minutes. Add garlic, apples and curry spice blend. Cook, stirring frequently, for 5 minutes. Stir in vinegar, sugar crystals and salt. Simmer, stirring occasionally, for 15 minutes or until apples and bell pepper are very soft and most of the liquid has evaporated. Set aside to cool.
2. Add chopped walnuts and stir to combine well. Transfer to an airtight container and store in the refrigerator for up to 1 week.

Chicken Satay with Peanut Sauce

Makes 8 servings

This satay is exceptionally tasty and easy to prepare. I often double or triple the recipe and serve them at room temperature for cocktail parties or with pre-dinner drinks. They're always a hit.

2 skinless boneless chicken breasts
2 cups Peanut Papaya Sauce (page 160) or Shiitake Peanut Sauce (page 159), divided

1. Preheat barbeque or broiler to high.
2. Using a meat pounder or rolling pin, pound chicken between 2 sheets of parchment paper until flattened to less than 1/2 inch. Cut each breast lengthwise into 4 strips and transfer to a shallow dish. Drizzle with 1 cup of peanut sauce. Cover and refrigerate for at least 1 hour or overnight.
3. Thread each strip of chicken onto an oiled skewer. Grill, basting with marinade, for 3 minutes on each side or until cooked through and chicken reaches 165°F on a meat thermometer. Discard marinade. Serve with remaining 1 cup peanut sauce.

You can use wooden skewers, but metal cooks the chicken faster because it conducts heat. Soak wooden skewers for at least 30 minutes before using.

Coconut Caponata

Makes 3 cups

Caponata is a traditional Italian dish featuring eggplant, garlic and tomatoes, often served with antipasti or as a relish. Toasted almonds and coconut give this version an Oceanic flair. Use it as an accompaniment for cooked meat, poultry and fish, as a sauce for cooked rice or pasta, or on toasted bread slices for a Polynesian bruschetta. Basil makes a great garnish.

- 3 tbsp melted coconut oil
- 1 onion, chopped
- 3 cloves garlic, minced
- 1 small eggplant, diced
- 3 tomatoes, peeled and chopped
- 1-1/4 cups Coconut Gremolata (page 149)
- 1/2 cup chopped pitted black olives (optional)

1. In a saucepan, heat oil over medium heat. Sauté onion for 5 minutes. Add garlic and eggplant and cook, stirring frequently, for 3 minutes or until eggplant begins to turn golden brown. Add tomatoes and cook, stirring occasionally, for 15 minutes or until eggplant is soft, the liquid from the tomatoes has been reduced and the mixture has thickened. Stir in gremolata and olives (if using) and heat through.

Coconut Shrimp

Makes 16 to 20 shrimp

Because coconut shrimp is one of my favourite restaurant appetizers, I wanted to find an easy way to make these tasty morsels without a deep fryer. This recipe looks a bit more complicated than it actually is. Kids love an assembly-type dish because it gets them involved in the kitchen, so turn them loose! You can do this with 1/4-inch slices of carrot, squash, sweet potato or sweet bell peppers.

- 1 lb (500 g) large cooked peeled shrimp, tail on (about 16 to 20 shrimp, deveined)
- 1/2 cup white wine
- 2 tbsp coconut aminos or tamari sauce
- 2/3 cup cornstarch
- 2 large eggs
- 1 cup unsweetened coconut flakes, fresh or dried
- 1/4 cup melted coconut oil

1. Using a knife, make 2 or 3 small incisions on the inside of each shrimp to prevent them from curling once cooked. Rinse, drain and pat dry.
2. In a bowl, combine wine and aminos. Add shrimp, stir to coat well and set aside, stirring occasionally, for 30 minutes.
3. Meanwhile, set up an assembly line to coat and fry the shrimp: Place cornstarch in a resealable bag. In a small bowl, beat eggs. Spread coconut flakes over the bottom of a shallow plate.
4. Drain shrimp (discard marinade). Transfer to a bowl.
5. Heat oil in a large, deep skillet over medium heat. Toss 3 to 6 shrimp in cornstarch to coat well (number of shrimp you coat and cook at one time depends on how many you can fit in your skillet without touching). Working one at a time, holding shrimp by the tail, coat with egg first and then dredge in coconut flakes. Drop prepared shrimp into skillet, spacing well apart from each other, and fry for about 1 minute or until coating on the bottom turns golden. Turn shrimp over and fry for 30 seconds or until coating is golden.
6. Using tongs, transfer cooked shrimp to a shallow dish lined with paper towel. Using a mesh skimmer or slotted metal spoon, skim off and discard the browned bits from the oil. Repeat Step 5 until all of the shrimp have been cooked.

You can marinate the shrimp for up to 2 hours in the refrigerator before coating and cooking them, but do not marinate for longer or the texture will begin to break down.

It is easier to use fresh cooked shrimp, but you can purchase frozen cooked shrimp and thaw them overnight in the refrigerator. Be sure to rinse the thawed shrimp under cool running water, then drain and pat dry before marinating in Step 2.

If using uncooked shrimp, increase the amount of oil so that it covers the shrimp while they cook. Shrimp are done when they turn bright pink.

CAUTION: If the amount of cooking oil is increased, you must use a deep pot or deep fryer to cook the shrimp in order to avoid serious splattering and burns.

Coconut Mango Salsa

Makes 3-1/2 cups

This salsa is big and bold and great with grilled fish or seafood, chicken or pork. The clear tastes of mango and coconut are enhanced by the shallot and cilantro (I added flat-leaf parsley as an option for cilantro-haters, like me).

- 3 ripe mangoes, diced
- 1 shallot, finely chopped
- 1 cup freshly grated coconut or reconstituted dried unsweetened coconut flakes
- 1/4 cup chopped fresh cilantro or flat-leaf parsley
- 2 tbsp freshly squeezed lemon juice
- 2 tbsp coconut nectar
- 1 tbsp melted coconut oil

1. In a bowl, combine mangoes, shallot, grated coconut, cilantro, lemon juice, nectar and oil and toss to mix well. Store in an airtight container in the refrigerator for up to 6 hours.

Because it is raw, this healthy salsa is best served as soon as it's made. The liquids do tend to separate upon standing. If this happens, strain using a fine-mesh sieve.

It's best to use freshly grated coconut in this recipe, but if you don't have fresh coconut on hand, soak dried shredded unsweetened coconut in 1/4 cup water while you prepare the vegetables. Drain (saving the soaking water for other recipes).

Toasted Coconut Chips

Makes 2 cups

Kids love these tasty snacking chips. Moms and dads love them because they aren't deep-fried and they don't have all sorts of chemicals added. Everybody wins with these simple bites. You can mix them with toasted nuts for a twist.

1/2 lb (250 g) fresh coconut meat
1 tsp Polynesian Spice Blend (page 140) or Garam Masala Spice Blend (page 137)
Sea salt (optional)

1. Preheat oven to 325°F.
2. Using a mandolin or vegetable peeler, slice coconut meat into wafer-thin, 2-inch-long strips. Arrange in a single layer on a rimmed baking sheet. Sprinkle with spice blend and season with salt (if using).
3. Bake in preheated oven, stirring occasionally, for 7 to 10 minutes or until coconut is crisp and golden brown. Set aside to cool. Store in an airtight container at room temperature for up to 3 days or in the refrigerator for up to 5 days.

This is one of those recipes that depend on your stove to determine the correct cooking time. I work with a convection oven that cooks much faster than a regular oven, so check the coconut often and write down the time that is optimal for you.

This recipe works best using fresh coconut meat—see page 12 for directions on extracting fresh coconut from the shell. If you have to use dried coconut, look for wide chips or flakes, but decrease the cooking time and watch carefully so that the coconut does not burn.

For a really healthy snack, toss 1 cup shredded kale with 1 tbsp melted coconut oil and the coconut strips before spreading on the baking sheet. Proceed with the rest of the instructions.

Vegetable Rice Stacks

Makes 4 servings

Want a stunning opener for a special dinner? This is it! It's delicious, healthy, easy to prepare, with no fussing over plating, yet it makes a fabulous visual statement. Add even more interest by garnishing with freshly grated flakes of coconut or sprigs of fresh thyme or oregano.

- 2 tbsp grated creamed coconut
- 1/4 cup coconut water or plain water
- 2 tbsp melted coconut oil
- 2 cloves garlic, finely chopped
- 1 tbsp finely chopped gingerroot
- 3 cups shredded green cabbage
- 3 cups shredded red cabbage
- 3 cups chopped or shredded bok choy
- 1/2 cup coconut water or chicken broth
- 3 tbsp butter
- 3 tbsp chopped raisins
- Sea salt and freshly ground black pepper
- Eight 3- × 3-inch rice wonton wrappers or fresh pasta sheets
- Sea salt and freshly ground black pepper

1. In a bowl, combine creamed coconut and 1/4 cup coconut water. Set aside.
2. In a skillet or wok, heat oil over medium heat. Sauté garlic and ginger for 1 to 2 minutes or until soft. Add green and red cabbage and bok choy and stir to coat well. Add 1/2 cup coconut water and bring to a boil. Add creamed coconut mixture. Reduce heat to low and simmer, stirring occasionally, for 15 minutes or until cabbage is tender-crisp and most of the liquid has evaporated. Stir in butter and raisins. Season to taste with salt and pepper.
3. Meanwhile, bring a pot of salted water to a boil. Add the wonton wrappers and cook according to package instructions. Using a colander, rinse with cool running water and drain well.
4. To serve, spoon 2 heaping tablespoons of the cabbage mixture onto each of 4 serving plates. Top with a wonton wrapper. Spoon 2 tablespoons of cabbage over each wrapper. Place 1 more wonton wrapper on each serving and top with a final layer of cabbage.

Creamed coconut is available in the baking ingredients section of well-stocked supermarkets (see page 14 for more information on this versatile product).

You can find fresh, pre-rolled sheets of pasta in the refrigerated section of most supermarkets. If you need to cut the sheets down to size, cook them first (in Step 3) and cut them after they cool.

Turkish Koftas with Peanut Sauce

Makes 4 appetizer servings

Once you make these spicy morsels, you'll see how easy they are. The skewers make them a great finger food for parties, but you can also roll them and bake them as meatballs. Serve with peanut sauce, as suggested here, or Banana Coconut Dip (page 132).

1/4 lb (125 g) ground turkey or chicken
1/4 lb (125 g) ground lean lamb
2 green onions, white and green parts, finely chopped
1 clove garlic, minced
1 tbsp Garam Masala Spice Blend (page 137)
1/4 tsp ground cayenne pepper
1 tbsp chopped fresh mint
Pinch of sea salt
1 cup Peanut Papaya Sauce (page 160) or Shiitake Peanut Sauce (page 159)

1. Set an oven rack to the highest position and turn oven to broil.
2. In a bowl, combine turkey, lamb, green onions, garlic, spice blend, cayenne, mint and salt. Using your hands, mix well. Divide mixture into 8 even portions and shape each into a small egg-shaped ball. Thread 2 balls onto each skewer and press each ball gently to form an elongated 1/2- to 3/4-inch flattened disc. Transfer to a platter, cover tightly and chill for at least 30 minutes or up to 48 hours. Bring to room temperature before cooking.
3. Broil the koftas, turning once or twice, for 5 to 8 minutes or until cooked through and koftas reach 160°F on a meat thermometer. Serve with peanut sauce.

You can use wooden skewers or toothpicks for grilling, but I recommend that you invest in a set of stainless steel entrée- and cocktail-size skewers. Metal skewers do not require pre-soaking and transfer heat more uniformly than wood.

Ground meat tends to slip around on an oiled skewer, so for this recipe, do not oil the skewers.

Banana Coconut Dip

Makes 1 cup

Coconut and banana combine to make this refreshing dip. Serve with nachos and crudités. It also doubles as a tropical accompaniment to grilled fish and chicken.

- 2 tbsp melted coconut oil
- 2 shallots, finely chopped
- 2 cloves garlic, minced
- 2 tsp Polynesian Spice Blend (page 140) or Garam Masala Spice Blend (page 137)
- 1 tbsp finely chopped gingerroot
- 1/4 cup freshly squeezed orange juice
- 2 tbsp freshly squeezed lemon juice
- 1 can (5.4 oz/160 mL) coconut cream
- 2 bananas, cut into 1-inch chunks

1. In a saucepan, heat oil over medium heat. Sauté shallot and garlic for 3 minutes. Add spice blend and ginger and cook, stirring constantly, for 1 minute or until onions are soft. Add orange juice, lemon juice and cream and bring to a boil. Add bananas, reduce heat and simmer, stirring frequently, for 15 minutes or until sauce has thickened.

Shallots (*Allium ascalonicum*) are edible bulbs with a thin paper-like outer skin similar to onions. Shallots are smaller than onions and separate into cloves like garlic when peeled (in recipes, the sum of these "cloves" is considered a head or shallot). Shallots are sweeter and milder than most cooking onions and are often preferred over onions in sauce and dip recipes. If you don't have shallots, you can substitute 1/4 cup finely chopped sweet onion (such as Vidalia or Spanish onion) or yellow cooking onion. Here are some of the shallot varieties I have found at farmers' markets in the fall.

Longer and usually without separate cloves, banana shallots (at left) are easy to slice using a mandolin or food processor. Yellow shallot (at right) is probably the most common. You will find it in grocery stores and supermarkets, but support local growers in the fall, when their crops are abundant. Almost as common as the yellow shallot are many different red varieties, including this one (in middle), which is called Prisma.

6

Spice Blends, Sauces & Dressings

Spice Blends

Sauces and Dressings

Coconut Jerk Spice Rub

Makes 1/2 cup

Chipotle chile pepper and Tellicherry peppercorns give this typically Jamaican spice blend zing. Use it sparingly at first and add more once your taste buds adapt to the heat.

- 2 tbsp Tellicherry peppercorns
- 1 star anise
- 1/4 cup coconut sugar crystals
- 2 tbsp sea salt
- 1 tbsp ground chipotle chile pepper
- 1 tbsp ground cinnamon
- 1 tbsp ground allspice

1. Using a mortar and pestle or small electric grinder, pound or grind peppercorns and star anise to a coarse or fine powder as desired.
2. In a bowl, combine sugar crystals, salt, chipotle, cinnamon, allspice and ground peppercorns and anise. Transfer to an airtight glass jar with a lid, label and store in a cool, dark, dry place for up to 3 months.

Tellicherry peppercorns have a rich, mature flavour because they are left on the vine to develop.

Chipotle chiles (*Capsicum*) are hot chile peppers (usually jalapenos) that are smoked to give them a distinct flavour.

Glass is best for storing spices and spice blends. If you can't find dark-coloured glass jars, use clear glass and wrap a piece of paper around the jar to block the light (use it as a label).

COCONUT
JERK RUB

GARAM
MASALA

Garam Masala Spice Blend

Makes 1/2 cup

This pleasing, complex blend of flavours is naturally sweetened by the cinnamon, star anise and nutmeg, and the cardamom seeds add a nice zip. It's not hot-spicy, but it does impart a rich, addictive flavour to dishes. I usually use 1 tablespoon of garam masala in recipes that call for it, but you can start with 1 or 2 teaspoons and gradually add more as your taste buds adjust to the pure, fresh spices.

4 cloves
One 3-inch stick cinnamon, crushed
1 star anise, crushed
1 nutmeg, crushed
4 tbsp coriander seeds
2 tbsp cumin seeds
1 tbsp black peppercorns
1 tsp cardamom seeds

1. In a small dry skillet or spice wok, combine cloves, cinnamon, star anise, nutmeg, coriander, cumin, peppercorns and cardamom. Toast over medium heat, stirring frequently, for 3 to 4 minutes or until lightly browned and fragrant. Watch carefully and remove from the heat just as the seeds begin to pop or the spices will smoke and burn. Set aside to cool.
2. Using a mortar and pestle or small electric grinder, pound or grind the toasted spices to your desired consistency. Transfer to an airtight glass jar, label and store in a cool, dark, dry place for up to 3 months.

Crush whole spices with a mortar and pestle or using a rolling pin.

Hot Spice Paste

Makes 1/3 cup

This spice blend is one of the hottest in this chapter, so start by cooking with small amounts and add more as your taste buds adapt.

- 4 cloves garlic
- 2 fresh or dried cayenne peppers
- One 1/2-inch piece peeled gingerroot
- One 1/2-inch piece peeled fresh turmeric or 1 tbsp dried turmeric
- One 1/2-inch piece peeled fresh galangal (optional)
- 1 to 2 tbsp softened coconut oil
- 1 tbsp coconut sugar crystals
- 1 tsp sea salt

1. Using a mortar and pestle or small electric food processor, pound or chop garlic, cayenne, ginger, turmeric and galangal (if using) to a paste. Add 1 tbsp oil, sugar crystals and salt and blend to a soft paste, adding more oil as necessary to keep the paste moist. Store in an airtight glass or porcelain container in the refrigerator for up to 10 days.

Spice pastes are both easy to make and easy to add to recipes, but they must be stored in the refrigerator and they do not keep as long as dried spice blends. You can turn any of the spice blends in this chapter into a paste by blending them with soft or melted coconut oil.

Galangal (*Alpina galanga*) and turmeric (*Curcuma longa*) are the edible roots of plants in the ginger family. Galangal is hotter and more peppery in taste than ginger. Turmeric is slightly smaller in diameter than gingerroot and the flesh is bright orange. Fresh turmeric adds a slight citrus flavour along with bitter warmth. You can find the fresh roots of both galangal and turmeric in the spring and fall at Asian markets. Stock up and freeze them for use later in the seasons.

Madras Curry Spice Blend

Makes 1/2 cup

Curry is a blend of spices, not just one spice, and everyone who makes their own has a secret curry combination. Feel free to experiment with the spices and amounts used in this recipe to make it yours. The heat in this curry blend comes from the mustard, peppercorns and chiles—and the fresher they are, the hotter they will be. If you like the flavour but not the heat, reduce the amounts of those spices in this blend to suit your taste.

- 2 tbsp fenugreek seeds
- 2 tbsp coriander seeds
- 1 tbsp allspice berries
- 1 tbsp mustard seeds
- 1 tsp black peppercorns
- One 4-inch stick cinnamon, crushed
- 10 cardamom pods
- 5 dried chiles
- 10 dried curry leaves (optional)
- 2 tbsp ground turmeric

1. In a small, dry skillet or spice wok, combine fenugreek, coriander, allspice, mustard, peppercorns, cinnamon, cardamom and chiles. Toast over medium heat, stirring frequently, for 3 to 4 minutes or until lightly browned and fragrant. Watch carefully and remove from the heat just as the seeds begin to pop or the spices will smoke and burn. Set aside to cool.
2. Using a mortar and pestle or small electric grinder, pound or grind toasted spices with curry leaves (if using) until coarsely or finely ground, as desired. Mix in turmeric. Transfer to an airtight glass jar, label and store in a cool, dark, dry place for up to 3 months.

Fenugreek (*Trigonella foenum-graecum*), a plant in the legume family, is the one spice closely associated with the flavour of curry, so it is almost always included in curry blends. However, the herb known as "curry plant" (*Helichrysum italicum*), which exudes a faint odour of curry in the leaves, is often confused with curry spice blend, which, if authentically Indian, does not contain it.

If you don't grow your own and can't find dried curry leaves, simply omit them; they are not essential.

Polynesian Spice Blend

Makes 1/2 cup

Coconut sugar and toasted coconut flakes combine with sweet spices for a spicy-sweet "island" blend. The nice thing about this combination is that it works in both sweet and savoury dishes. It will round out your spice cabinet nicely.

- 1/4 cup toasted coconut flakes, fresh or dried
- 1/4 cup coconut sugar crystals
- 1 tbsp ground cinnamon
- 1 tsp ground allspice
- 1 tsp sea salt
- 1/2 tsp ground cloves

1. Using a mortar and pestle or small electric grinder, pound or grind toasted coconut until coarsely or finely ground, as desired. Transfer to a bowl and add sugar crystals, cinnamon, allspice, salt and cloves. Mix well. Transfer to an airtight glass jar, label and store in a cool, dark, dry place for up to 3 months.

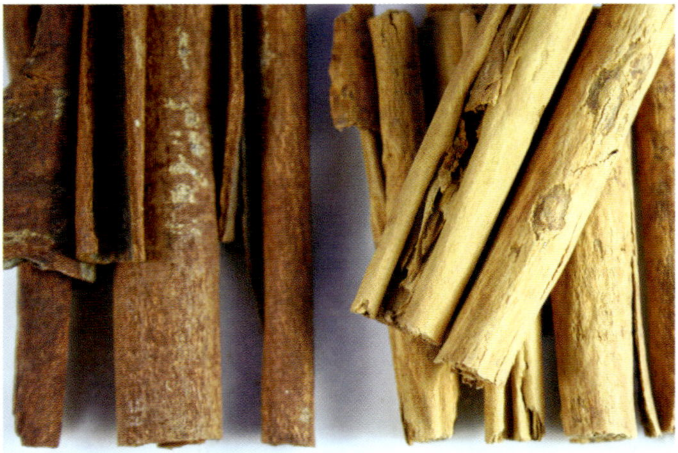

I use cinnamon as a sweet flavouring in both sweet and savoury dishes. It imparts an aromatic "island" taste to dishes. My favourite ground or stick cinnamon is premium Vietnamese cinnamon, but *Cinnamomum zeylanicum* (shown at right in the photograph) comes a close second. It is much sweeter than the common *Cinnamomum cassia* (shown at left in the photograph), with a pure, never-bitter cinnamon flavour.

You can find Vietnamese cinnamon at fine spice stores or online.

• POLYNESIAN •

Tandoori Spice Blend

Makes 3/4 cup

The smokey paprika is the wild card in this blend, giving it an exotic flavour that is hard to identify, while ginger, cayenne pepper and mustard provide the heat. Use this blend with caution to start, then adapt it to your tastes in future dishes.

3 tbsp smokey paprika
2 tbsp ground ginger
2 tbsp ground turmeric
1 tbsp ground cayenne pepper
1 tbsp ground coriander
1 tbsp ground cumin
1 tbsp ground mustard
1 tbsp sea salt
1/2 tsp ground cloves

1. In a bowl, combine paprika, ginger, turmeric, cayenne, coriander, cumin, mustard, salt and cloves. Mix well. Transfer to an airtight, dark-coloured glass jar, label and store in a cool, dry place for up to 3 months.

Vietnamese Five-Spice Blend

Makes 1/2 cup

If you are wondering about the name of this blend because you have counted the ingredients and come up with six, not five, you need to know that I don't count coconut sugar as a spice. This blend is different from the traditional Chinese five-spice blend because of its sweetness, which comes from the coconut sugar as well as the star anise and Vietnamese cinnamon.

- 2 tbsp ground coriander
- 2 tbsp ground star anise
- 2 tbsp coconut sugar crystals
- 2 tbsp ground Vietnamese cinnamon
- 1 tbsp ground fennel seeds
- 1 tbsp sea salt

1. In a bowl, combine coriander, star anise, sugar crystals, cinnamon, fennel and salt. Mix well. Transfer to an airtight, dark-coloured glass jar, label and store in a cool, dry place for up to 3 months.

You can find Vietnamese cinnamon at fine spice stores or online.

Coconut Dressing

Makes about 1-1/2 cups

Pure coconut flavour explodes in this easy-to-make dressing. I find it works best with fruit and summer greens, but it sure is nice to drift back to the tropics in the middle of winter with just a shake of a jar.

- 1 cup coconut milk
- 1/4 cup unsweetened shredded coconut, fresh or dried
- 3 tbsp melted coconut oil
- 2 tbsp coconut vinegar
- 1 tbsp Dijon mustard

1. In a jar, combine milk, shredded coconut, oil, vinegar and mustard. Cap and shake well to emulsify. Use immediately or cover and store in the refrigerator for up to 3 days. Bring to room temperature before using.

Coconut Blue Cheese Dressing

Makes about 2 cups

The amount of coconut cream or milk called for in the recipe tends to make a thinner dressing that can be drizzled over lettuce wedges and used as a sauce for barbequed chicken, vegetables or steaks. If you prefer a thicker dressing, simply reduce the amount of coconut cream or milk.

- 2 oz blue cheese such as Stilton
- Juice of 1/2 lemon
- 2 tbsp softened coconut oil
- 2/3 cup coconut cream or coconut milk
- 1 tsp coconut aminos or tamari sauce

1. In a bowl, mash the cheese using a fork. Stir in lemon juice and oil. Add the cream and aminos and mash until small pieces of the cheese are distributed in the dressing. Use immediately or cover and store in the refrigerator for up to 3 days. Bring to room temperature before using.

Honey Lime Coconut Dressing

Makes 1 cup

A tart-sweet and brilliantly crafted dressing.

- 1/3 cup freshly squeezed lime juice
- 1/4 cup coconut nectar or liquid honey
- 3 tbsp melted coconut oil
- 1 tbsp toasted sesame oil
- 1 tbsp tahini
- 1 tbsp tamari or soy sauce
- 1 tbsp coconut vinegar
- 1/4 tsp grated fresh gingerroot

1. In a jar, combine lime juice, nectar, coconut oil, sesame oil, tahini, tamari, vinegar and ginger. Shake well to emulsify. Use immediately or cover and store in the refrigerator for up to 6 days. Bring to room temperature and shake well before using.

Tahini is a paste made from ground sesame seeds. Depending on the brand, it may have oil floating on the top, which is easy to stir back into the paste before measuring. You may need to mash the tahini it with a fork before shaking the ingredients in order to get it incorporated evenly in the dressing.

Hot Spiced Coconut Sauce

Makes 1-1/2 cups

The spice paste used in this recipe is hot and slightly sweet, making this a versatile sauce for pasta or rice, vegetables or cooked meats, chicken or fish. Depending on your tolerance for spicy-hot seasoning, add more or less of the paste, to taste, or use 1 tbsp of any one of the milder spice blends found in this chapter.

- 3 tbsp melted coconut oil
- 1 small onion or 2 shallots, finely chopped
- 2 tbsp coconut flour
- 2 tbsp Hot Spice Paste (page 138) or 1 tsp hot sauce
- 1 can (14 oz/398 mL) coconut milk

1. In a saucepan, heat oil over medium heat. Add onion and cook, stirring constantly, for 2 minutes. Add flour and spice paste and cook, stirring constantly, for 1 or 2 minutes or until fragrant and lightly browned. Whisk in milk and cook, stirring constantly, for 3 minutes or until sauce has thickened. Use immediately or cover and refrigerate for up to 3 days.

Coconut Gremolata

Makes 1 cup

Gremolata is a savoury mixture of parsley and garlic that is used to garnish slow-cooked veal or lamb and fish as well as soup and other vegetable dishes. My version is nutty and slightly sweeter because I've replaced the garlic with coconut. If desired, you can add finely chopped fresh garlic to make it savoury.

- 1/2 cup chopped fresh parsley
- 1/4 cup toasted slivered almonds, coarsely chopped
- 1/4 cup toasted, coconut flakes, fresh or dried
- 1 tsp sea salt

1. In a bowl, combine parsley, almonds, coconut flakes and salt. Toss to mix well. Store in an airtight container in the refrigerator for up to 2 weeks or freeze for up to 1 month. Use as a garnish for soup, stew, salads and vegetable dishes.

To toast coconut: Preheat oven to 350°F. Arrange coconut in an even layer on a rimmed baking sheet and toast, stirring once or twice, for 5 to 7 minutes or until lightly brown.

I use a mezzaluna (above) to chop fresh herbs. The all-in-one, stainless steel handles on this one are easy to keep clean and do not harbour bacteria as do wooden handles.

To toast nuts or seeds, in a dry skillet or saucepan, heat nuts over medium-high heat, stirring frequently, for 2 to 3 minutes . . .

. . . or until golden brown. Remove from heat, transfer to a bowl and let cool. Do not let the nuts cool in the hot pan or they will continue to brown.

Coconut Tomato Sauce

Makes about 3 cups

Once you make your own tomato sauce, you'll find it hard to use a commercial one—the flavour is just that good. In the summer and early fall, when fresh, ripe, home-grown or local tomatoes are abundant, you will be able to quarter them and remove the skin easily using a paring knife. In the winter, use 1 can (28 oz/ 796 mL) of diced tomatoes and their juices in place of the fresh tomatoes.

- 2 tbsp melted coconut oil
- 1 onion, chopped
- 3 to 4 cloves garlic, finely chopped
- 1 tbsp Polynesian Spice Blend (page 140) or Garam Masala Spice Blend (page 137)
- 6 tomatoes, peeled and chopped
- 1 can (5.4 oz/160 mL) coconut cream
- 10 fresh basil leaves, chopped
- 1 tbsp chopped fresh oregano
- 1/2 tsp sea salt

1. In a skillet, heat oil over medium heat. Sauté onion for 5 minutes. Add garlic and spice blend and cook, stirring frequently, for 2 minutes or until garlic is soft. Add tomatoes and cream and bring to a boil. Reduce heat and simmer for 10 minutes or until reduced. Add basil, oregano and salt and cook, stirring, for 1 minute or until heated through. Use immediately or cover and refrigerate for up to 3 days.

If the tomatoes are not perfectly ripe, you will need to blanch them to remove the skins.

Using the tip of a paring knife, core and then make an "x" on the bottom of each tomato. Drop the tomatoes into boiling water and boil for 1 minute.

Using a slotted spoon, lift the tomatoes out of the boiling water and plunge into a bowl of ice water. Let cool, drain and slip the skins off.

The tomatoes are now ready to be sliced or chopped.

Orange Coconut Dressing

Makes 1-1/2 cups

Citrus juices are great in dressings—as the acid, they add a brilliant tone to the oil-acid mix. With coconut in the blend, the citrus is muted, and the dressing becomes a subtle, creamy sauce worthy of all kinds of raw and cooked fruit and vegetables.

- 1 tbsp grated orange rind
- Juice of 1 orange
- Juice of 1 lemon
- 1/2 banana, cut into chunks
- 1 can (5.4 oz/160 mL) coconut cream
- 3 tbsp melted coconut oil
- 1 tbsp tahini

1. In a blender, combine orange rind, orange juice, lemon juice, banana, cream, oil and tahini. Blend for 30 seconds or until creamy. Use immediately or cover and refrigerate for up to 6 days. Bring to room temperature and stir well before using.

Tahini is a paste made from ground sesame seeds. Depending on the brand, it may have oil floating on the top, which is easy to stir back into the paste before measuring. You may need to mash the tahini with a fork before adding it to the blender in order for it to incorporate evenly in the dressing.

Sesame Dressing

Makes 1 cup

I'm partial to sesame in part for its medicinal value as a source of vitamin E and coenzyme Q10, but mostly for its nutty flavour. In this cooked dressing, the sesame sings.

- 2 tbsp melted coconut oil
- 2 cloves garlic, minced
- 1 tbsp freshly grated gingerroot
- 1 tbsp coconut flour
- 1 cup chicken broth
- 1 tbsp coconut aminos or tamari sauce
- 1 tbsp tahini

1. In a saucepan, heat oil over medium heat. Add garlic and ginger and cook, stirring frequently, for 2 minutes. Add flour and stir into a paste. Whisk in broth and aminos and cook, stirring frequently, for 3 to 5 minutes or until sauce has thickened. Remove from heat and stir in tahini. Use immediately or cover and refrigerate for up to 6 days. Bring to room temperature and stir well before using.

Tahini is a paste made from ground sesame seeds. Depending on the brand, it may have oil floating on the top, which is easy to stir back into the paste before measuring. You may need to mash the tahini with a fork before adding it to the saucepan in order for it to incorporate evenly in the dressing.

Sweet and Sour Pineapple Sauce

Makes 2 cups

For such a simple sauce, this adds a tangy finish to dishes. I've paired it with Sweet Potato Shrimp Cakes (page 90), but it is so versatile that if I have it in the refrigerator, I use it in many other dishes, from chicken and fish to vegetables.

- 2 tbsp melted coconut oil
- 1 onion, finely chopped
- 3 cloves garlic, finely chopped
- 1 can (14 oz/398 mL) crushed pineapple and juices
- 1/2 cup coconut sugar crystals
- 1/4 cup coconut vinegar
- 1 ripe banana, mashed
- Juice of 1 lime

1. In a saucepan, heat oil over medium heat. Sauté onion for 5 minutes or until soft. Add garlic and cook, stirring frequently, for 2 minutes. Stir in pineapple and juices, sugar crystals, vinegar, banana and lime juice. Bring to a boil, reduce heat and simmer for 20 to 30 minutes or until liquid has reduced and sauce has thickened. Use immediately or cover and refrigerate for up to 6 days. Bring to room temperature and stir well before using.

Coconut Garlic Mayonnaise

Makes 1 cup

Making your own mayonnaise has definite health advantages, and once you've tried the homemade version, you may never go back to store-bought. Serve as a dip or spread for sandwiches or add it to recipes that require mayo.

- 1/2 cup melted coconut oil
- 1/2 cup avocado oil or olive oil, at room temperature
- 1 clove garlic
- 1 large egg yolk
- 1 large egg
- 1 tbsp freshly squeezed lemon juice
- 1 tbsp Dijon mustard
- 1 tsp sea salt

1. In a glass measuring cup or bowl, whisk together coconut oil and avocado oil.
2. In a blender, chop garlic. Add egg yolk, egg, lemon juice, mustard and salt. Blend for 30 seconds or until the mixture reaches the consistency of a smooth paste. With the motor running, slowly pour the oil mixture through the opening in the lid. Blend until well combined. Use immediately or transfer to an airtight container and refrigerate for up to 3 days. (The mayonnaise will thicken as it chills.)

CAUTION: This is an additive-free mayonnaise made with raw eggs, which means that it must be refrigerated immediately after making and will only keep in the refrigerator for up to 3 days.

Avoid using homemade mayonnaise in picnic sandwiches or salads where refrigeration is unreliable or in dips or sauces that will sit out at room temperature for longer than 30 minutes.

To reduce the risk of bacteria from the shell being transferred to the inside, gently wash eggs in soapy water and store them in a clean plastic or ceramic egg keeper.

For a spiced mayonnaise: In Step 2, add 2 tsp of any of the spice blends found in this chapter in place of or in addition to the mustard.

For a Caesar-style dressing: Add 1 drained anchovy with the flavourings in Step 2.

For a thinner salad dressing: Whisk in 1 can (5.4 oz/160 mL) coconut cream in Step 2.

Coconut Yogurt Mayonnaise

Makes 2-1/2 cups

Adding yogurt to mayonnaise doubles the yield and reduces the calories per serving. Strained yogurt is similar to Greek-style yogurt and is often called "yogurt cheese" because it is thick and creamy, almost like creamed cheese.

- 1/2 cup melted coconut oil
- 1/2 cup avocado oil or olive oil
- 1 large egg yolk
- 1 large egg
- 1 tbsp coconut vinegar
- 1 tsp sea salt
- 1 can (5.4 oz/160 mL) coconut cream
- 1 cup strained plain low-fat yogurt (see Recipe Note)

1. In a glass measuring cup or bowl, whisk together coconut oil and avocado oil.
2. In a blender, combine egg yolk, egg, vinegar and salt. Blend for 30 seconds or until it reaches the consistency of a smooth paste. With the motor running, slowly pour the prepared oil mixture through the opening in the lid. Blend until well combined. Transfer to a bowl.
3. Whisk in cream. Using a spatula, fold in strained yogurt. Use immediately or transfer to an airtight container and refrigerate for up to 3 days. (The mayonnaise will thicken as it chills.)

CAUTION: This is an additive-free mayonnaise made with raw eggs, which means that it must be refrigerated immediately after making and will only keep in the refrigerator for up to 3 days.

Avoid using homemade mayonnaise in picnic sandwiches or salads where refrigeration is unreliable or in dips or sauces that will sit out at room temperature for longer than 30 minutes.

To reduce the risk of bacteria from the shell being transferred to the inside, gently wash eggs in soapy water and store them in a clean plastic or ceramic egg keeper.

To strain yogurt: Line a sieve with a double layer of cheesecloth and place the sieve over a bowl. Empty a 1-lb (500 g) container of plain low-fat yogurt into the sieve and place it in the refrigerator to drain for up to 2 hours. The longer the yogurt drains, the thicker the yogurt becomes. Reserve the drained liquid and use it to replace milk or water in smoothies, soup, stews and baked goods.

Shiitake Peanut Sauce

Makes 3 cups

This peanut sauce is very different from most: the coconut milk makes it light and creamy; the sweet spice blend tempers the peanuts; and the shiitake mushrooms give it body and an exotic, earthy flavour—a must try.

- 2 tbsp melted coconut oil
- 1 onion, finely chopped
- 3 cloves garlic, minced
- 1-1/2 cups sliced shiitake mushrooms
- 1 tbsp Polynesian Spice Blend (page 140) or Garam Masala Spice Blend (page 137)
- 1-1/2 cups roasted salted peanuts
- 1 tbsp creamed coconut
- 1/4 cup coconut milk

1. In a saucepan, heat oil over medium heat. Sauté onion for 5 minutes. Add garlic, mushrooms and spice blend and cook, stirring frequently, for 5 minutes or until mushrooms release their liquid and are tender. Remove pan from heat and set aside to cool.
2. In a small food processor or blender, combine onion-mushroom mixture and any pan juices, peanuts and creamed coconut and process for 30 seconds. With motor running, add milk through opening in the lid. Pulse or process until you reach desired consistency. Use immediately or transfer to an airtight container and refrigerate for up to 3 days.

I like to make my own peanut sauce because I can make it smooth, thin, thick or crunchy, depending on how I plan to use it. Sometimes I coarsely chop the peanuts separately and add to the sauce after everything is blended, but you can leave them whole and add them to the sauce at the end of Step 2.

Try using raw or toasted cashews or almonds in place of the peanuts.

Peanut Papaya Sauce

Makes 2 cups

I like to use crunchy-style peanut butter to make this peanut sauce, but use smooth peanut butter if you prefer. It can be a bit difficult to combine the broth with the peanut butter, so use a fork or an electric mixer, but it's not critical if it isn't completely blended.

- 1/4 cup peanut butter
- 1/3 cup chicken or vegetable broth
- 2 tbsp melted coconut oil
- 4 cloves garlic, finely chopped
- 4 green onions, white and green parts, finely chopped
- One 1-inch piece peeled gingerroot, finely chopped
- 1 tbsp Polynesian Spice Blend (page 140) or Garam Masala Spice Blend (page 137)
- 1 can (5.4 oz/160 mL) coconut cream
- 2 fresh papayas, peeled, seeded and chopped

1. In a glass measuring cup or bowl, combine peanut butter and broth.
2. In a saucepan, heat oil over medium heat. Add garlic and onions and sauté for 2 minutes or until garlic is soft and slightly golden. Add ginger and spice blend and cook, stirring constantly, for 1 minute. Stir in coconut cream and the peanut butter mixture and bring to a boil. Add papayas, reduce heat and simmer for 5 minutes or until mixture is thick and papayas have blended into the sauce. Use immediately or transfer to an airtight container and refrigerate for up to 3 days.

You can substitute any fresh, soft-fleshed fruit—peaches, nectarines, plums, mangoes—for the papayas or you can omit the fruit altogether.

7

Beverages

Berry Booster

Makes 2 drinks

Antioxidants from the vitamin E found in wheat germ and the blueberries along with the fatty acids from the oil make this a fine immune booster and anti-aging drink.

- 1/2 cup coconut water
- 2 tbsp melted coconut oil
- 1 tbsp wheat germ
- 1 cup blueberries, fresh or frozen
- 1 cup sliced strawberries, fresh or frozen

1. In a blender, combine coconut water, oil, wheat germ, blueberries and strawberries. Secure lid and blend until smooth.

If you're new to using coconut oil, you may wish to start with 1 teaspoon of coconut oil and gradually increase to 2 tablespoons over time.

Energizer Smoothie

Makes 2 drinks

This smoothie will keep you going through the mid-morning or mid-afternoon blahs. The fibre from the fruit and wheat germ combined with the protein in the yogurt, peanut butter and coconut oil will give you energy and satisfy your appetite so your mind doesn't wander.

- 1/2 cup coconut water
- 1/2 cup yogurt
- 1 to 2 tbsp melted coconut oil
- 2 tbsp wheat germ
- 2 tbsp smooth peanut butter
- 2 tbsp coconut nectar
- 2 bananas, cut into chunks
- 1 cup frozen strawberries

1. In a blender, combine coconut water, yogurt, oil, wheat germ, peanut butter, nectar, bananas and strawberries. Secure lid and blend until smooth.

If you're new to using coconut oil, you may wish to start with 1 teaspoon of coconut oil and gradually increase to 2 tablespoons over time.

You can use plain or flavoured yogurt, but be sure to check that it contains active bacterial cultures.

Hawaiian Antioxidant Punch

Makes 10 drinks

This is a great drink to have on hand for an after-school or after-sport thirst quencher.

3 cups freshly squeezed orange juice, divided
Juice of 1 lime
4 mangoes, coarsely chopped
1 can (14 oz/398 mL) crushed pineapple and juices
4 cups chilled coconut water
2 cups crushed ice

1. In a blender, combine 1 cup orange juice, lime juice, mangoes and pineapple with juices. Secure lid and blend until smooth.
2. Transfer mixture to a punch bowl. Stir in coconut water and remaining orange juice and keep chilled until ready to serve. Add ice just before serving.

To keep your punch cold longer, place all of the ingredients along with the punch bowl in the refrigerator for an hour before serving (or overnight).

Minted Papaya Quencher

Makes 4 drinks

Mint is an antispasmodic, so it relaxes muscles and helps to reduce stiffness and aches after strenuous workouts. If this drink is consumed about an hour before meals, the digestive enzymes in the pineapple, papaya and mint significantly aid digestion and keep tummies bloat-free.

- 1 cup coconut water
- 1/3 cup pineapple juice
- 1/2 cup crushed ice
- 1 tbsp freshly squeezed lime juice
- 2 papayas, flesh and seeds
- 1 banana, cut into chunks
- 5 sprigs fresh peppermint, divided

1. In a blender, combine coconut water, pineapple juice, ice, lime juice, papayas and banana. Strip the leaves from 1 sprig of peppermint and add to the blender. Secure lid and blend until smooth. Divide among 4 glasses and garnish each with a sprig of peppermint.

Immune Builder

Makes 2 drinks

Fortify your defenses with this delicious drink filled with beta carotene and vitamin C, which increase the production of white blood cells and antibodies. By fighting germs and relieving stress on the immune system, coconut oil allows white blood cells to function more efficiently. When a cold or flu first threatens, start taking this drink with echinacea at least four or five times a day.

- 1/2 cup coconut milk
- 1 to 2 tbsp coconut oil
- 1/3 cup ground almonds
- 2 mangoes, cubed
- 1 cup cubed cantaloupe

1. In a blender, combine milk, oil, almonds, mangoes and cantaloupe. Secure lid and blend until smooth.

If you're new to using coconut oil, you may wish to start with 1 teaspoon of coconut oil and gradually increase to 2 tablespoons over time.

Peppermint Tea

Makes 1 drink

This is my summer pick-me-up drink. Mint is naturally uplifting, and when blended with sweet cicely and coconut oil, the taste is delicious.

- 2 small sprigs fresh peppermint
- 1 tbsp chopped fresh sweet cicely or lemon balm
- 1 tbsp softened coconut oil

Sweet cicely (*Myrrhis odorata*) is a sweet-tasting herb that lends a mild anise flavour to tea blends. (It is well worth planting in your herb garden.) You can substitute lemon balm (*Melissa officinalis*) or bergamot (*Mondara didyma*) for the sweet cicely.

1. Strip the leaves from the sprigs of peppermint and combine with sweet cicely in a mug. Pour in boiling water, stir and let steep for 5 to 7 minutes. Remove leaves, if desired. Stir in oil.

You can substitute 1 tablespoon dried peppermint for the fresh peppermint.

If you're new to using coconut oil, you may wish to start with 1 teaspoon of coconut oil and gradually increase to 2 tablespoons over time.

Oil and water do not mix on their own without emulsifiers, so coconut oil tends to float to the top of teas. Leave the spoon in your cup and stir the tea each time you take a sip.

Watermelon Elixir

Makes 4 drinks

This drink is more than a great summer cooler; it's also full of nutrients. Watermelon contains vitamins A, B and C and lycopene, which are essential for helping to protect against a growing list of cancers.

- 1/4 cup Polynesian Spice Blend (page 140)
- 2 tbsp sea salt
- 1 lime, cut into quarters
- 1 watermelon, seeded and cubed (about 6 cups)
- 1/4 cup coconut cream or coconut milk
- 2 tbsp grated creamed coconut

1. In a saucer or shallow dish, combine spice blend and salt. Run the lime quarters around the rims of 4 glasses (reserve the lime quarters). Turn glasses upside down and twist the rims into the spice blend to coat, or rim, them. Set aside.
2. In a blender, combine watermelon, cream and creamed coconut. Secure lid and blend until smooth. Pour into prepared glasses and garnish each with reserved lime quarter.

8

Desserts

Banana Cake with Mango Icing

Makes one 9-inch double-layer cake

This could easily elbow out my 25-year-old daughter's all-time favourite chocolate mocha cake to become her birthday cake tradition. The tropical flavours of banana and mango sent her over the moon when I tested this recipe. Her dad loves it, too, so it's likely going to be a twice-annual celebration cake.

- 1-1/2 cups mashed ripe bananas
- 1 tsp pure vanilla extract
- 1/4 cup coconut cream or coconut milk
- 1-1/4 cups all-purpose flour
- 1 cup coconut flour
- 2 tsp baking powder
- 3/4 tsp baking soda
- 1/2 tsp sea salt
- 1/4 cup softened coconut oil
- 1/4 cup unsalted butter, at room temperature
- 1-1/2 cups coconut sugar crystals
- 2 large eggs
- 2 cups Mango Icing (page 174)

1. Preheat oven to 350°F. Line the bottom of two 9-inch round cake pans with parchment paper.
2. In a bowl, combine bananas, vanilla and cream. Set aside.
3. In a separate bowl, whisk together all-purpose flour, coconut flour, baking powder, baking soda and salt. Set aside.
4. In a mixing bowl or the bowl of a stand mixer, beat together the oil and butter until light and creamy. Add sugar crystals slowly, beating until well blended. Add eggs, one at a time, and continue to beat until fluffy.
5. In alternating batches, add one-third of the dry ingredients and one-third of the banana mixture, mixing well after each addition, until a smooth batter has formed.
6. Pour equal amounts of the batter into prepared cake pans, tapping the sides of the pans so that the batter is evenly distributed. Bake in preheated oven for 30 minutes or until a toothpick inserted into the centre of the cake layers comes out clean and the layers begin to pull away slightly from the sides of the pans. Invert cake layers onto wire racks and set aside to cool completely.
7. Meanwhile, prepare Mango Icing.
8. Transfer 1 cake layer to a plate. Spread reserved mango purée overtop (see Mango Icing, Step 2, page 174). Top with the remaining cake layer. Spread icing evenly over the top and sides of the cake.

You will need 4 to 5 very ripe bananas for this cake. You can mash them using a fork or potato masher, but for this cake, I prefer a smooth purée, so I process them in a blender or food processor.

For both the cake batter and icing, be sure not to liquefy the coconut oil. This means you will have to watch the oil as it softens in the pan of hot water and remove it as soon as it softens to the consistency of room-temperature butter.

Coconut sugar crystals replace granulated sugar in the cake batter. Look for raw, enzymatically active, organic coconut sugar crystals—they are low glycemic and contain amino acids, minerals, vitamin C and broad-spectrum B vitamins.

Mango Icing

Makes 2 cups

If you make just the mango purée (to the end of Step 2), it is great as a creamy fruit layer in fruit parfait or as an accompaniment to fresh fruit. With the addition of cream cheese and icing sugar to the purée, this recipe becomes a cool twist on traditional cream cheese icing.

- 1/3 cup coconut water or plain water
- 1/3 cup granulated sugar
- 2 tbsp freshly squeezed lemon juice
- 3 mangoes, chopped
- 1/4 cup spreadable cream cheese
- 1/4 cup softened coconut oil
- 3 to 4 cups icing sugar

1. In a saucepan, combine water, granulated sugar and lemon juice. Bring to a boil over medium heat. Reduce heat and simmer, stirring constantly, for 2 minutes. Add mangoes and cook, stirring occasionally, for 12 minutes or until soft. Set aside to cool completely.
2. Using a blender or potato masher, combine the mango mixture with the cream cheese to make a smooth purée. Reserve one-third of the mango purée for cake filling.
3. Add oil to the remaining mango purée and mix well. In 1-cup batches, add icing sugar, stirring well after each addition to make a soft, spreadable icing.

Baked Peaches with Coconut Streusel

Makes 4 servings

Coconut stars in every part of this exceptional dessert. The streusel topping gives a nutty, crunchy finish to the baked peaches.

- 1/3 cup softened coconut oil
- 1/2 cup coconut sugar crystals
- 1/3 cup coconut flour
- 1/4 cup large-flake rolled oats
- 1/4 cup coarsely chopped almonds
- 1/2 tsp ground cinnamon
- 1/8 tsp sea salt
- 4 peaches
- 1/2 cup coconut cream or Whipped Coconut Cream Topping (optional; page 192)
- 1/3 cup toasted shredded coconut, fresh or dried

1. Preheat oven to 375°F. Lightly oil a 9- × 13-inch baking dish.
2. In a bowl, using a wooden spoon, cream oil and sugar crystals. Add flour, oats, almonds, cinnamon and salt and beat until well combined. Mixture will be crumbly and coarse in texture. Set aside.
3. Cut peaches in half and remove pits. Arrange peaches, cut side up, in a single layer in prepared baking dish. Divide streusel into 8 equal portions and press evenly on top of each peach half. Bake for 15 minutes or until streusel has browned. (Be careful not to overcook or peaches will fall apart.)
4. Serve warm. Arrange 2 peach halves on each of 4 dessert plates. Drizzle with cream or top with a dollop of whipped topping (if using). Garnish with toasted coconut.

For this recipe, large just-ripe freestone peaches or nectarines are best. Freestone varieties are firm and less juicy than clingstone varieties, so it is easier to remove the pit and keep the peach halves in one piece. My favourite freestone varieties are Red Haven and Redstone; other varieties include Glowhaven and Cresthaven (at left).

To toast coconut: Preheat oven to 350°F. Arrange coconut in an even layer on a rimmed baking sheet and toast, stirring once or twice, in preheated oven for 5 to 7 minutes or until lightly brown.

Brazilian Quindins

Makes 12 quindins

A popular Brazilian dessert, these little cakes or puddings are called quindins. My version uses more egg white, which makes them slightly lighter than those typically found in Brazil. The coconut floats to the top while cooking and leaves a jelly-like layer of custardy sugar and egg at the bottom. Remove cooked cakes from the pan and turn upside down so the coconut acts as a crust for the custard, which should be on the top when served.

- 5 large eggs
- 1/2 cup coconut sugar crystals
- 1/4 cup lightly packed brown sugar
- 1 tbsp softened coconut oil or butter
- 2 cups sweetened coconut flakes, fresh or dried

1. Preheat oven to 350°F. Lightly grease a 12-cup muffin pan.
2. Separate eggs, placing 5 yolks in a bowl and 3 whites in a separate bowl. Discard or reserve remaining 2 egg whites for another use. Using a hand-held or electric mixer, beat egg whites until stiff peaks form. Set aside.
3. In a bowl, blend sugar crystals, brown sugar and oil. Beat in the egg yolks. Stir in coconut flakes. Fold in the beaten egg whites. Spoon into prepared muffin pan (at left).
4. Prepare a *bain-marie*: Place muffin pan in a larger baking pan filled with 1-inch boiling water. Bake for 35 minutes or until firm and golden. Remove pan from *bain-marie* and set aside to cool until the sides of the cakes begin to pull away from the sides of the pan. Transfer to a cooling rack, flipping the quindins so the coconut is on the bottom and the egg custard is on the top.

A *bain-marie* (hot-water bath) is used to gently cook the egg and sugar mixture into a "custard" and prevents the eggs from cooking too fast and hardening like scrambled eggs.

Coconut Cherry Oat Squares

Makes 20 squares

Coconut flour is gluten-free and high in protein as well as fibre, but it does take some time to adjust to the way it performs in baked goods. It absorbs liquids; therefore, recipes featuring coconut flour require more eggs and other liquids. Baked goods made with coconut flour are also more crumbly than those made from wheat flour (due to the lack of gluten).

- 6 large eggs
- 3 cups coconut sugar crystals
- 1 tbsp pure vanilla extract
- 3 cups sweetened shredded coconut, fresh or dried
- 1/2 cup dried cherries
- 1/2 cup large-flake rolled oats
- 6 tbsp coconut flour
- 1 tsp sea salt

1. Preheat oven to 350°F. Lightly oil a 9- × 13-inch baking pan.
2. In a bowl, using a hand-held or electric mixer, beat eggs. Add sugar crystals and vanilla and beat until smooth. Stir in shredded coconut, cherries, oats, flour and salt. Press into the prepared pan and bake for 15 to 20 minutes or until set and golden.
3. Transfer to a wire rack to cool for 10 minutes. Cut into 2-inch squares and transfer to a plate to cool completely. Squares keep in an airtight container for 3 days at room temperature and up to 5 days in the refrigerator.

Coconut Cream Pie

Makes 1 pie

What can I say? It's richly flavoured with coconut, it's silky smooth and it melts in your mouth. It's also easy to make. I think you'll agree that this is a blue-ribbon dessert.

- 1/2 cup coconut sugar crystals
- 1/4 cup cornstarch
- 2 cups coconut milk
- 1 tsp pure vanilla extract
- 1/2 cup sweetened shredded coconut, fresh or dried
- 1 Coconut Pie Crust (page 182), baked and cooled
- 2-1/2 cups Whipped Coconut Cream Topping (page 192)
- 1/4 cup toasted shredded coconut, fresh or dried

1. In a saucepan, combine sugar crystals and cornstarch. Whisk in milk and vanilla. Bring to a boil, add shredded coconut and cook, stirring constantly, for 2 to 5 minutes or until thickened. Set aside to cool.
2. Scrape into prepared pie crust and spread evenly. Cover and refrigerate until ready to serve. Spread whipped topping evenly over pie and sprinkle with toasted coconut.

To toast coconut: Preheat oven to 350°F. Arrange coconut in an even layer on a rimmed baking sheet and toast, stirring once or twice, in preheated oven for 5 to 7 minutes or until lightly brown.

VARIATION: Turn this into a Banana Cream Pie by slicing 2 ripe bananas into the bottom of the baked coconut pie crust just before adding the filling in Step 2.

Coconut Pie Crust

Makes 1 pie crust

Try using this coconut pie crust in place of the traditional pie crust in all of your favourite pie recipes—it adds a lightly sweet and nutty flavour that's just brilliant. (Some recipes in this chapter, such as Chocolate Coconut Cheesecakes on page 183, call for this crust unbaked.)

- 2-1/2 cups sweetened shredded coconut, fresh or dried
- 1 cup finely chopped pecans
- 2/3 cup coconut sugar crystals
- 2 large egg whites
- 3 tbsp melted coconut oil
- 1/2 tsp sea salt

1. Preheat oven to 300°F.
2. In a bowl, combine shredded coconut, pecans, sugar crystals, egg whites, oil and salt.
3. Press into a 9-inch pie plate and bake in preheated oven for 15 to 20 minutes or until lightly golden. Set aside to cool.

TO MELT CHOCOLATE: Coarsely chop the chocolate and place in the top of a double boiler or heatproof bowl (not plastic) set over a pan of simmering water. (The bottom of the bowl should be about an inch above the water so that only the steam heats it. Make sure the bowl is dry inside and be careful not to splash any water into the chocolate—even the smallest amount of water will cause chocolate to lose its smooth and creamy texture and cause it to seize and clump.) Stir frequently and remove the bowl from the heat when the chocolate has melted. Dry the bottom of the bowl and transfer it to a wire rack to cool. Do not add hot chocolate to the cheese mixture—it will cook the eggs and affect the texture of the cheesecakes. Once melted, chocolate will stay liquid for about an hour, but it will harden if left in a cool place or left overnight, so melt the chocolate before starting the recipe in order to allow time for it to cool.

Chocolate Coconut Cheesecakes

Makes 6 individual cheesecakes

The nice thing about these sensational cheesecakes is that you can make them in custard cups of any size to suit the occasion. If you bake them in wide-mouth canning jars, you can put the lids on, tie bows around them and take them to a special event as an exceptional dessert contribution.

- 6 oz dark chocolate
- 1 Coconut Pie Crust (page 182), unbaked
- 170 g cream cheese, softened
- 113 g ricotta cheese
- 1/2 cup coconut sugar crystals, ground until fine
- 2 large eggs
- 1 tsp pure vanilla extract
- 1/2 cup Coconut Fudge Sauce (page 184)
- 1 cup Whipped Coconut Cream Topping (optional; page 192)

1. Preheat oven to 300°F.
2. Melt and cool chocolate (see facing page).
3. Arrange 6 custard cups or ramekins (3-1/2 to 4-inch diameter) on a baking sheet. Divide pie crust into 6 equal portions and, using the back of a spoon, press each portion into the bottom of a custard cup. Bake in preheated oven for 15 to 20 minutes or until light golden. Transfer to a cooling rack and set aside to cool. Leave oven on.
4. In a food processor, combine cream cheese, ricotta cheese, sugar crystals, chocolate, eggs and vanilla and process for 1 minute or until smooth. Spoon filling into shells and bake for 20 to 25 minutes or until set (a knife inserted into the centre comes out clean). Let cool. Cover and refrigerate for at least 1 hour or overnight.
5. Serve chilled or at room temperature. Drizzle with fudge sauce and top with a dollop of whipped topping (if using).

You can bake the crust in the custard cups and refrigerate up to 48 hours in advance. Bring to room temperature before filling and baking the cheesecakes.

For the antioxidant benefit, use dark chocolate with 70% or more cocoa solids.

To grind coconut sugar crystals: Using a spice grinder or clean coffee grinder, process coconut sugar until fine.

Coconut Fudge Sauce

Makes 1-1/2 cups

It's coconut, it's fudgy—it's good. Very good.

- 2 cans (5.4 oz/160 mL each) coconut cream
- 1 cup coconut sugar crystals
- 1 cup brown sugar
- 1 tsp sea salt

1. In a saucepan, combine cream, sugar crystals, brown sugar and salt. Bring to a boil over medium-high heat, stirring constantly, until sugars are melted. Reduce heat and simmer, without stirring, for 7 minutes.
2. Remove from heat, stir and pour into a buttered 9-inch pie plate or 9- × 9-inch baking dish. Set aside to cool. Transfer to a sauceboat or bowl and cover tightly. Store in refrigerator for up to 5 days. Serve as a topping for ice cream, cake, custard and other desserts.

The sauce will thicken on cooling but may separate as the coconut oil solidifies. If it separates, warm it (by partially immersing the container in hot water, not boiling) and stir with a fork before using.

Coconut Strawberry "Ice Cream"

Makes 4 to 6 servings

This recipe is adapted from a recipe by Sidra Forman that appeared in *Edible DC* magazine. I love it because it's a delicious vegan alternative to ice cream. You'll need an ice-cream maker to make this recipe.

2-1/2 cups coconut milk
1/2 cup fresh coconut water and immature coconut meat (optional)
1-1/2 cups halved fresh strawberries or 1 cup puréed strawberries
1/4 cup + 2 tbsp softened coconut oil
1/2 cup coconut nectar
1 tsp sea salt

1. In a blender, combine milk, coconut water and immature coconut meat (if using), strawberries, oil, nectar and salt. Secure lid and blend until smooth. Taste and add more nectar if needed.
2. Freeze, following manufacturer's instructions for your ice-cream maker.

You can substitute raw coconut water for the coconut meat if a fresh, immature coconut is not available.

VARIATION: You can substitute 1 cup of any puréed fruit for the strawberries.

Easy Coconut Lemon Tarts

Makes 24 tarts

Of course, for lemon lovers, no recipe collection would be complete without a tart lemon dessert. This is an uncooked version that sets thanks to the coconut. The texture is definitely not silky smooth: the coconut adds significant bite to the filling.

- 1/2 recipe Coconut Pie Crust (page 182), unbaked
- 1/2 cup coconut milk
- 4 tbsp melted coconut oil
- 3 tbsp coconut nectar
- 2 cups sweetened shredded coconut, fresh or dried
- 1 packet (7 oz/200 g) creamed coconut, cut into chunks
- Grated rind from 2 lemons
- 1/2 cup freshly squeezed lemon juice
- 1/2 tsp sea salt
- 1 cup Whipped Coconut Cream Topping (optional; page 192)

1. Preheat oven to 300°F.
2. Fill each well of two 12-cup tart pans with about 1 tablespoon of coconut pie crust, pressing crust evenly up the sides. Bake in preheated oven for 15 to 20 minutes or until lightly golden. Set aside to cool.
3. In a blender or food processor, combine milk, oil, nectar, shredded coconut, creamed coconut, lemon rind, lemon juice and salt. Secure lid and process until combined.
4. Scoop about 1 tablespoon of the filling into prepared tart shells. Cover and refrigerate for at least 1 hour or overnight. Serve chilled or at room temperature with a dollop of whipped topping (if using).

Macaroons

Makes 18 macaroons

The egg whites are not beaten to stiff peaks in this recipe. One bowl, a fork, a tablespoon and a baking tray are all the equipment you need to prepare these classic coconut cookies. Be sure that the oven is preheated before you bake them.

- 2 large egg whites
- 1/2 cup coconut sugar crystals
- 2 cups medium grated coconut, fresh or dried

1. Preheat oven to 350°F. Line a baking sheet with parchment paper.
2. In a bowl, using a fork, beat the egg whites for about 20 seconds or until frothy. Beat in sugar crystals. Stir in grated coconut, 1 cup at a time, mixing with the fork in between batches to make sure there are no dry spots in the mixture.
3. Using a tablespoon and pressing each scoop against the side of the bowl to compact it, drop mounds of coconut batter onto prepared baking sheet. Flatten slightly with the fork. Bake for 10 to 15 minutes or until the base of the macaroons is golden. Transfer to a cooling rack and let cool. Store in an airtight container at room temperature for up to 3 days or in the refrigerator for up to 7 days.

Using sweetened coconut in this recipe makes the macaroons almost like candy, so I prefer to use unsweetened coconut.

Prepare these delicious treats as large or as small as you wish. To make 18 macaroons, I use a standard measuring tablespoon.

Because they are coconut-dense, macaroons do not spread out as they bake, so you can arrange them fairly close together on the baking sheet.

VARIATION: For something different, try pressing a whole almond (I prefer salted toasted almonds) into the centre of each macaroon before baking. You can also dip one half of each cooled macaroon in melted dark chocolate.

Easy Chocolate Tarts

Makes 18 tarts

I'm not a dessert person. While I did have a heck of a sweet tooth growing up, I don't usually eat pie or cake after a meal. But I do like to serve dessert when people come for dinner, and this is such an easy one to make, especially because you can freeze the leftovers.

- 1/2 recipe Coconut Pie Crust (page 182), unbaked
- 2 cans (5.4 oz/160 mL each) coconut cream
- 10 oz dark chocolate, chopped
- 1 cup Whipped Coconut Cream Topping (optional; page 192)

1. Preheat oven to 300°F.
2. Fill each well of a 12-cup tart pan and six wells of a second tart pan with about 1 tablespoon of coconut pie crust, pressing crust evenly up the sides. Bake in preheated oven for 15 to 20 minutes or until lightly golden. Set aside to cool.
3. In a saucepan, bring cream to a simmer over medium heat. Add chocolate and cook, stirring constantly, for 2 minutes or until chocolate has melted. Set aside to cool.
4. Pour chocolate coconut cream filling into prepared tart shells. Freeze for 10 minutes or until the filling is set. Serve chilled or at room temperature with a dollop of whipped topping (if using).

This is a rich dessert and a tiny tart is ample for most people, so use a tart tin (not a muffin tin) with 1- × 1-inch wells.

For the antioxidant benefit of chocolate, use dark chocolate with 70% or more cocoa solids.

Oatmeal Cookies

Makes 12 cookies

The mango and coconut chips take this cookie from an ordinary, healthy oatmeal treat to a soft and chewy dessert experience. Make the cookies small or large, but make them often for kids and adults alike.

- 1/2 cup coconut flour
- 1/2 cup all-purpose flour
- 1-1/2 cups large-flake rolled oats
- 1/2 tsp baking powder
- 1/2 tsp baking soda
- 1/4 tsp sea salt
- 1 tsp Polynesian Spice Blend (page 140) or ground cinnamon
- 1/2 cup coconut milk or coconut water
- 1 large egg
- 1 ripe banana
- 1/4 cup coconut nectar or pure maple syrup
- 1/4 cup melted coconut oil
- 1 tsp pure vanilla extract
- 1 mango, diced
- 1/4 cup sweetened coconut chips or flakes, fresh or dried

1. Preheat oven to 350°F. Line a baking sheet with parchment paper.
2. In a large bowl, combine coconut flour, all-purpose flour, oats, baking powder, baking soda, salt and spice blend.
3. In a separate bowl, using a fork, beat milk and egg until combined. Mash banana into milk mixture. Add nectar, oil and vanilla and mix well.
4. Make a well in the dry ingredients. Add the wet ingredients to the dry ingredients and stir until well combined. Gently fold in the mango and coconut chips until evenly distributed. Cover and refrigerate for 5 to 10 minutes.
5. Using a 4-ounce cookie scoop or 2 heaping tablespoons, drop dough on prepared baking sheet, spacing each about 1 to 2 inches apart. Flatten slightly with a fork. Bake in preheated oven for 15 minutes or until golden on the bottom and firm on top. Let cool on baking sheet for 2 to 3 minutes or until set and transfer to a wire rack to cool completely. Store in an airtight container at room temperature for up to 5 days or freeze for up to 3 months.

Whipped Coconut Cream Topping

Makes 2-1/2 cups

The calories in this whipped topping are high—about 200 per serving—but the fat from the coconut is more desirable than the additives from non-dairy toppings. For best results, chill cans of coconut cream for at least 3 hours or overnight.

- 2 tbsp coconut sugar crystals
- 1 tbsp coconut flour
- 2 cans (5.4 oz/160 mL each) coconut cream, chilled
- 1 tsp pure vanilla extract

1. In a blender or small food processor, process sugar crystals to a fine powder. Transfer to a bowl and stir in flour.
2. In a deep bowl, combine coconut cream and vanilla. Using an electric mixer, beat until thick and fluffy. Fold in powdered sugar and flour mixture, a little at a time, until well blended and thick. Use immediately or cover and refrigerate for up to 3 days.

The quality of the coconut cream is important in this recipe. There may be other brands that work, but I have used Native Forest with great success. If you can't find a good-quality coconut cream, use the following substitution.

To substitute for 2 cans of coconut cream: Chill 2 cans (14 oz/398 mL each) coconut milk for at least 30 minutes. Flip the cans upside down and open the bottom of 1 can. Scoop out the solid "cream" and measure 3/4 cup (you may need to open the second can). Store the remaining liquid in an airtight container in the refrigerator for up to 3 days and use in recipes that call for coconut milk.

9

Spa Recipes

Coconut Hand Lotion

Makes 1/4 cup (about 24 applications)

When first made, this lotion is very thin, but you can still use it right away. Over time, the coconut oil will set at room temperature, causing the lotion to thicken slightly and become more spreadable. Use about 1/2 teaspoon for each application.

- 3 tbsp softened coconut oil
- 1 tbsp almond oil
- 2 tsp glycerine

1. In a bowl, whisk together coconut oil and almond oil. Add glycerine and whisk for several minutes or until well blended. Transfer to a jar with a lid (1/3 cup capacity), cap, label and store in a cool place.

Vitamin E is usually added to homemade creams and lotions as an antioxidant and preservative. However, because the recipes in this book are made using coconut oil, which is stable at room temperature, there is no need to add vitamin E.

Coconut Body Oil

Makes a generous 1/2 cup (about 4 applications)

Coconut's antimicrobial properties come from its medium-chain fatty acids and lauric acid. These components make it an excellent ingredient for body creams and lotions. What I particularly like about using coconut oil is its ability to absorb completely into my skin within minutes.

1/4 cup melted coconut oil
1/4 cup almond oil
3 tbsp olive oil
Few drops perfumed oil

1. In a small jar with a lid (3/4 cup capacity), combine coconut oil, almond oil and olive oil. Add perfumed oil, a few drops at a time, until the fragrance of the body oil is to your liking. Cap, label and shake vigorously. Store in a cool, dark place . Shake well before using liberally after a bath or shower.

This recipe combines 2 polyunsaturated oils that will go rancid in heat and after long exposure to light, so make small quantities and use within 2 weeks.

Because all of the ingredients in face, hand and body creams are absorbed through the pores of the skin, it's important to use organic, virgin, expeller-pressed oils to make them—in other words, as high quality as you use in cooking.

Look for unrefined organic almond oil in specialty health or alternative food stores. You can use organic virgin olive oil in place of the almond oil if desired.

Perfumed oil is derived from flowers, although it may not be "essential" oil. Jasmine, rose, orange and verbena are some of the fragrance oils you can find sold in small dropper bottles at specialty health or alternative food stores. Store this body oil in a small, dark-coloured jar if you have one. Keep it in a cool place, but not the refrigerator. If the coconut oil solidifies, cup the jar between your palms to warm it and soften the coconut oil.

Coconut Tanning Oil

Makes 1 cup

While health experts don't encourage overexposure to sunlight, I recognize that some people like the effects of the sun on their skin. Both coconut oil and sesame oil contain a natural screening component that absorbs harmful ultraviolet rays and helps to prevent cell damage from exposure to sunlight. The tea in the oil encourages the skin to darken more quickly and helps to reduce the exposure time needed to result in tanned skin.

- 1/2 cup boiling water
- 2 bags regular (black) tea
- 1/4 cup softened coconut oil
- 1/4 cup sesame oil
- 3 tbsp lanolin

1. In a small teapot, pour boiling water over tea bags. Cover and steep for 10 minutes. Remove and discard tea bags. Set tea aside to cool.
2. In the top of a double boiler or heatproof bowl, combine coconut oil, sesame oil and lanolin. Set over boiling water, being careful that the bottom of the double boiler or bowl does not touch the water. Heat, stirring constantly, for 1 or 2 minutes or until combined.
3. Remove from heat and slowly whisk in the prepared tea. Pour into a jar with a lid (1-1/4 cup capacity). Cap, label and store at room temperature. If oil separates, shake well before using.

CAUTION: Do not use on children. UVA and UVB rays from direct sunlight are linked to cancer. When your skin cells darken, this is a signal that they have already been exposed to and damaged by the sun.

Fair skin is more prone to skin damage and cancer than darker skin types. Know your skin type and how quickly it burns. Do not rely on this oil to prevent damage to your skin or to protect it against the cancerous effects caused by exposure to sunlight.

Face and Body Scrub

Makes 4 cups (several applications)

This body scrub not only exfoliates dead skin cells, it also moisturizes, leaving your skin feeling soft, with a delicate coconut fragrance.

- 2 cups salt or granulated sugar
- 2 cups unsweetened desiccated coconut, fresh or dried
- 1/2 cup softened coconut oil

1. In a blender, combine salt, coconut and oil. Blend for 1 minute or until a thick paste has formed.
2. Transfer to a jar with a lid (1-quart capacity) and refrigerate. Gently rub a liberal amount of scrub over your face, hands and body after showering. Rinse off and pat dry with a towel.

Depending on the grade, sea salt can melt into this mixture. Table salt works best in this recipe.

Herb and Coconut Deep Hair Conditioner

Makes scant 3/4 cup (about 12 applications)

This conditioner is easy to apply and makes hair soft and silky. Sometimes after deep conditioning for a couple of hours, I find that I need to shampoo my hair twice, so I will shampoo it after the initial application and then again the next morning. If you find the glycerine stays in your hair, rinse with coconut vinegar or apple cider vinegar.

- 1/2 cup melted coconut oil
- 2 tbsp jojoba oil
- 1 tbsp glycerine
- 1/4 tsp rosemary oil or chamomile oil

1. In a jar with a tight-fitting lid (3/4 cup capacity), combine coconut oil, jojoba oil, glycerine and rosemary oil. Cap, label and shake well. Store at room temperature and shake well before using.

You can find essential oils at health or alternative food stores. Use rosemary oil for dark hair and chamomile oil for light hair.

To use as a regular conditioner: Pour a dime-size portion into the palm of your hand and rub palms together. Apply to wet hair and comb through. Dry naturally or blow-dry and style as you normally do.

To use as a deep hair conditioner: Pour a quarter-size portion into the palm of your hand and rub palms together. Apply to dry hair and massage thoroughly into hair and scalp. Cover your hair with a plastic shower cap and allow the conditioner to penetrate the hair for a minimum of 1 hour or overnight.

Lavender Bath Vinegar

Makes 2 cups (1 application)

The coconut oil doesn't mix with the bathwater but instead floats on the surface. As you rise out of the water, it coats your body, leaving it moisturized and soft. Use any dried aromatic herb or a mixture of herbs in place of the lavender, including crushed dried bay leaves, sage, chamomile or lemon balm.

- 1/4 cup dried lavender buds
- 2 tbsp melted coconut oil
- 1 cup boiling water
- 1 cup coconut vinegar or apple cider vinegar

1. Place the lavender and oil in a 4-cup liquid measuring cup or quart Mason jar. Pour boiling water overtop. Cover and let steep for 30 minutes. Add vinegar, stir, cover and let steep at room temperature for 1 hour or overnight.
2. To use, draw the bathwater and, using a fine-mesh sieve, strain the vinegar into the tub as it fills with water.

If desired, double this recipe to make enough bath vinegar for 2 baths.

To store, strain into an open or loosely capped bottle (such as an apothecary glass jar with stopper). Vinegar will keep at room temperature for several months.

Lemon Balm Spritz

Makes 3 cups

Lemon balm (*Melissa officinalis*) is a calming herb with the taste and fragrance of lemon. It is used in teas to help ease tension, restlessness, nervous sleep disorders and digestive and gastrointestinal disorders. Its antibacterial and antiviral properties make it a good addition to lotions and cosmetic waters.

- 1/2 cup crushed dried lemon balm leaves
- 1/4 cup grated lemon rind
- 1 tbsp crushed allspice berries
- 1 cup coconut vinegar
- 2 cups coconut water or plain water

1. In a pint jar with a lid, combine lemon balm, lemon rind, allspice and vinegar. Shake well and set aside in a sunny window to steep for 10 days.
2. Using a fine-mesh sieve, strain vinegar (discarding herbs) into a jar with a lid (1-quart capacity) and add coconut water. Cap, label and shake well. Store in the refrigerator for up to 1 month.
3. To use, pour some of the spritz into a small atomizer bottle. Use daily as a refreshing face toner before applying makeup or throughout the day to refresh the skin.

CAUTION: Do not use lemon balm during pregnancy.

Turmeric Coconut Face Mask

Makes 1/4 cup (2 applications)

Turmeric (*Curcuma longa*) is a powerful anti-inflammatory, antimicrobial, antibacterial, antifungal and wound-healing herb, which makes it a perfect ingredient in skincare products. This combination of natural ingredients helps to reduce redness and dark age spots or dark circles under the eyes. For a relaxing weekly or monthly beauty ritual, take a long, hot bath using Lavender Bath Vinegar (page 201). Apply Face and Body Scrub (page 198) as directed. Spritz with Lemon Balm Spritz (page 202) and apply Turmeric Coconut Face Mask as directed below.

- 1 heaping tbsp ground turmeric
- 1 tbsp liquid raw honey
- 1 tbsp melted coconut oil
- 1 tbsp 2% milk

1. In a small glass bowl, add turmeric. Using a long-handled teaspoon, stir in honey. Add oil and milk and stir to make a smooth paste. Divide mixture in half. Use one half immediately and store the remaining mixture, tightly covered, in the refrigerator for up to 3 days.
2. To use, remove makeup, wash face and neck and pat dry. Using the back of a spoon, spread mask evenly over cheeks, forehead, chin, nose and under the eyes. Leave mask on for 15 to 20 minutes. Rinse off with warm water, pat dry, spritz with Lemon Balm Spritz (page 202) or toner and moisturize with coconut oil.

This mask does not dry and tighten the skin. When gently rinsed off with warm water, the lactic acid in the milk acts as an exfoliant and softens the skin. For best results, apply once or twice a week and use pure coconut oil as a daily non-greasy moisturizer.

CAUTION: Turmeric is a yellow dye that will stain, so use a long-handled spoon to mix and apply it and be sure to wear old clothes or an old apron when working with it. Be aware that it will stain facecloths and towels, ceramic bowls, as well as your porcelain sink. To remove yellow turmeric stains from porcelain, wipe the surface with the cut edge of half a lemon and rinse well.

ACKNOWLEDGEMENTS

I'm incredibly grateful to be able to do what I love, which is to explore gardens and experiment with wholesome, organic food and to write about those experiences. I'm also blessed with family and friends who share some of my passion for herbs and gardens, food markets, restaurants and, of course, healthy, fresh food. To all those people with whom I have shared a fine meal, thank you for joining me in one of life's most precious entitlements.

My talented daughter, Shannon McLauchlin, designed the spice labels shown on pages 135, 136 and 141.

I think what makes HarperCollins a great publisher is their understanding of the process of building books. They appreciate and give full rein to authors. Thanks to Brad Wilson, Alan Jones, Kelly Hope, Erin Parker, Kelly Jones, Ruth Pincoe, Natalie Garriga, Beverley Sotolov and Maria Golikova. Tracy Bordian trained her eye on the copy to tweak and tighten my words into cohesive text and I value her attention to detail.

The cover photographs and most of the recipe photographs were shot by Chris Campbell, styled by Dennis Wood and exquisitely propped by Susan Rogers—such a creative and food-savvy team. Thank you for the high-energy, productive work that you did to illustrate the beauty and versatility of coconut.

Developing and testing recipes not only takes time and talent, it also requires a lot of ingredients and equipment. I live in southern Ontario—surrounded by the rich farmland of Grey County—and I'm thrilled to say that we have a large number of energetic, like-minded young farmers who share a deep abiding love for the land, the plants and the animals. My heartfelt thanks to those who are restoring the land, acre by acre, while providing me with winter and summer organic vegetables, herbs and meat.

Like the passionate growers of organic food, some enlightened companies helped in no small way with the testing for *Coconut 24/7*. Coconut Secret shared their flour, sugar crystals, vinegar, nectar and aminos. Edward & Sons contributed exquisite-quality Native Forest coconut milk, coconut cream and creamed coconut—all organic. Spectrum Organic Products Inc. offered their expeller-pressed organic coconut oil. For 100% pure coconut water, I turned to Waiola. I am grateful to all for their generosity and help with product information, and I can say from first-hand use that these products are exceptional.

While reading an issue of *Edible DC*, I found a delightful recipe for Strawberry Coconut Ice Cream by Sidra Forman, who graciously granted permission to adapt her recipe for this book.

This book builds on the scientific work of many dedicated people who seek to enjoy truly nourishing food. Thanks to the fine work of Dr. Mary Enig and Sally Fallon (authors of *Eat Fat Lose Fat*); Bruce Fife (author of *The Coconut Oil Miracle*) and Siegfried Gursche (author of *Coconut Oil*).

Index

A

B

C

Q

R

S

T

V

W

Y

Z